H. Carrington Bolton

The Students' Guide in Quantitative Analysis

H. Carrington Bolton

The Students' Guide in Quantitative Analysis

ISBN/EAN: 9783337138639

Printed in Europe, USA, Canada, Australia, Japan

Cover: Foto ©berggeist007 / pixelio.de

More available books at **www.hansebooks.com**

THE

STUDENTS' GUIDE

IN

QUANTITATIVE ANALYSIS.

INTENDED AS AN AID TO THE STUDY OF

FRESENIUS' SYSTEM.

BY

H. CARRINGTON BOLTON, Ph.D.,

PROFESSOR OF CHEMISTRY IN TRINITY COLLEGE,
HARTFORD, CONN.

ILLUSTRATED.

THIRD EDITION, WITH ADDITIONS AND CORRECTIONS

SECOND THOUSAND

NEW YORK:

JOHN WILEY & SONS

1899.

Printed by
Braunworth, Munn & Barber,
Brooklyn, N. Y., U. S. A.

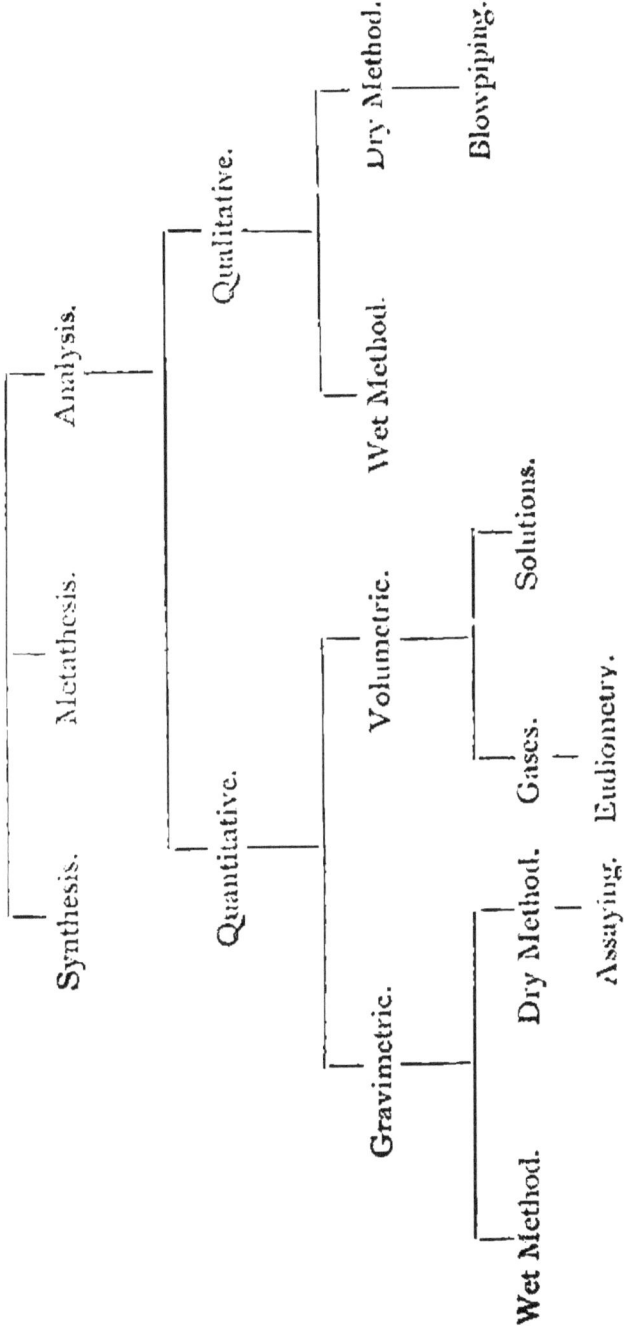

Synthesis.

Analysis.

Metathesis.

Quantitative.

Qualitative.

Gravimetric.

Volumetric.

Wet Method.

Dry Method.

Assaying.

Gases.

Eudiometry.

Solutions.

Wet Method.

Dry Method.

Blowpiping.

PREFACE.

A portion of the following pages originally appeared in the columns of the *American Chemist*, under the title: "*Schemes of Analyses executed in the School of Mines, Columbia College.*" Numerous applications for copies in book form have induced the author to publish the Schemes under a more general title.

Since writing the articles the author has been called to another sphere of labor, and the circumstances which led to their compilation are explained in the following paragraphs, quoted from the prefatory remarks accompanying the original publication.

"The system of instruction in Quantitative Analytical Chemistry, organized in the School of Mines, Columbia College, by Dr. C. F. Chandler, has been developed by the Assistants, who have had charge of the Laboratory for Quantitative Analysis, Mr. Alexis A. Julien, Dr. Paul Schweitzer, and the writer.

The practical examples and the methods of analysis were originally selected by Prof. Chandler; the latter have been modified by the Assistants, and from time to time they have introduced new processes, conforming to the advances made in this department of chemical science.

The plan of the STUDENTS' GUIDE is similar to that in the excellent papers of Mr. Alexis A. Julien entitled: "Examples for Practice in Quantitative Analysis," the details, however, are the result of observing the needs of students during my five years' experience in teaching large classes.

The fragmentary character of many portions of the notes is accounted for by the fact that they are intended to serve in part as lecture notes, and to indicate to the student the points to be studied. FRESENIUS' "*System of Instruction in Quantitative Chemical Analysis*" (American edition, by Profs. O.D. Allen and S. W. Johnson; New York, 1881) is placed in the hands of each student on entering the laboratory, but many students are perplexed by the peculiar, though systematic, arrangement of this classic work, and are at a loss to know *how* to begin work, *what to study*, and *where to find* the information appropriate to particular cases. To aid the student in the study of Fresenius' work, and not to displace it, is one of the objects of the STUDENTS' GUIDE. It is then scarcely necessary to state that very free use has been made of *Fresenius' System;* acknowledgment is, however, made in all cases. By occasional references to original papers the student's attention is directed to methods, as detailed by their authors, with the hope of encouraging the student in research."

H. C. B.

Trinity College.

LIST OF ANALYSES.

INTRODUCTORY NOTES.

By means of Chemical Analysis we determine the composition of any substance.

The object of Qualitative Analysis is to determine the *nature* of the constituents of a body.

The object of Quantitative Analysis is to determine the *amount* of these constituents.

Quantitative Analysis includes two methods, Gravimetric and Volumetric Analysis.

In Gravimetric Analysis we convert the known constituents of a compound into such forms as will admit of their exact determination by weight. This is done chiefly in two ways:

1st. By separating one of the constituents of a body as such (*e.g.*, Cu by the battery).

2nd. By converting an existing constituent into a new form by exchange of elements (*e.g.*, $AgNO_3 + HCl = AgCl + HNO_3$).

The forms must fulfil two conditions:

1st. Must be capable of being weighed exactly.

2nd. Must be of known and fixed composition.

The choice of form of precipitate depends on two consid-erations. The most preferable are—

1st. Those most insoluble in the surrounding liquid.

2nd. Those in which the proportion of the constituents to be determined is very small compared with the weight of the precipitate (*e.g.*, S in $BaSO_4$ is only 13.7 per cent.).

In Volumetric Analysis the amount of a constituent is estimated by the action of reagents in solutions of known strength and of determined volumes. (See Notes on Volumetric Analysis, p. 40).

Fresenius. A System of Instruction in Quantitative Chemical Analysis. *Editions:* American, John Wiley and Sons, New York, 1881 ; last English ; last German.

Thorpe. Quantitative Chemical Analysis. New York, last edition.

Rose, H. Traité Complet de Chimie Analytique. Paris, 1859-62. 2 vols.

Rose H., and *Finkener.* Handbuch der analytischen Chemie. Leipzig, 1867.

Mohr. Lehrbuch der chemisch-analytischen Titrirmethode. Braunschweig, last edition.

Sutton. Systematic Handbook of Volumetric Analysis. London, last edition.

Rammelsberg. Leitfaden für die quantitative chemische Analyse. Berlin, 1863.

Crookes. Select Methods in Chemical Analysis. London, last edition.

Bolley and *E. Kopp.* Handbuch der technisch-chemischen Untersuchungen. Leipzig, last edition.

Wöhler. Die Mineral Analyse in Beispielen. Göttingen, 1861. *Also translation* by Henry B. Nason. Philadelphia, 1871.

Prescott. Outlines of Proximate Organic Analysis. Van Nostrand, New York, last edition.

Caldwell. Agricultural Qualitative and Quantitative Chemical Analysis. New York, last edition.

Wanklyn. Water Analysis. London. last edition.

Bunsen. Anleitung zur Analyse der Aschen und Mineral-wasser. Heidelberg, 1874.

Ricketts. Notes on Assaying and Assay Schemes. New York, Wiley & Sons ; last edition.

Storer. First Outlines of a Dictionary of Solubilities of Chemical Substances. Cambridge, 1864.

Heppe. Die Chemische Reactionen der wichtigsten anor-ganischen und organischen Stoffe. (Tabellen, etc.) Leipzig, 1875.

Zeitschrift für Analytische Chemie, Fresenius. Wiesbaden, 1862 to date.

Jahresbericht über die Fortschritte der Chemie. Giessen, 1847 to date.

Bulletin de la Société Chimique de Paris. Paris, 1864 to date.

Chemical News. Crookes. London, 1860 to date.

American Chemist. Chandler. New York, 1870-77.

American Journal of Science and Art. J. D. and E. S. Dana. New Haven, 1819 to date.

Journal of Analytical Chemistry. Edward Hart. Easton, 1887 to date.

THE

STUDENTS' GUIDE

IN

QUANTITATIVE ANALYSIS.

Analysis No. 1.—BARIC CHLORIDE.

$BaCl_2 + 2H_2O.$

A.—Determination of Chlorine.

See Fres. Quant. Anal., § 141, I, a, and pages 790 to
795. (References are to Fresenius' *Quantitative Analysis,*
American edition, 1881.)

Weigh out 0.8 to 1 grm. of powdered $BaCl_2 + 2H_2O$ and
dissolve in cold water in a beaker; add a slight excess of
$AgNO_3$ previously acidulated with HNO_3; stir well, and
warm. When the precipitate of AgCl has entirely settled,
and the supernatant liquid is quite clear, pour off through
a No. 2 filter; then add boiling water slightly acidulated
with HNO_3, to the precipitate in the beaker; stir, and,
after the precipitate has settled again, pour off through
the filter. Continue this washing by decantation three or
four times; then bring the precipitate on the filter by
means of a glass rod or a feather; wash it down into the
point of the filter; wash lastly with a little non-acidified
water; cover the funnel with paper; label properly, and set
aside to dry. Weigh a clean porcelain crucible; transfer

the precipitate to this crucible, removing the AgCl from the paper as completely as possible. Wrap a clean platinum wire around the rolled-up filter, forming a "*cradle;*" burn the filter in the cradle over the inverted crucible cover ; do not let the ashes fall into the crucible. Moisten the ashes with conc. HNO_3 (one drop); heat one minute ; add a drop of conc. HCl; evaporate cautiously, and heat the contents of the crucible and cover until the AgCl is partly fused, avoiding carefully a higher temperature than necessary. See Fres., § 82, b. Weigh the crucible and contents. For calculation, see D.

AA.—SECOND METHOD.—Compare Fres., § 115, I, a, β. Take to 0.2 to 0.5 grm. $BaCl_2 + H_2O$; dissolve in warm water; acidulate with HNO_3 (free from chlorine); pour into a "parting flask;" add $AgNO_3$ in slight excess; cork the flask, and shake well. When well settled, wash the precipitate in the flask by decantation with warm water, without filtering. Invert the flask, covered with a watch-glass, over a weighed porcelain crucible, placed in a large porcelain dish, and filled with water. Withdraw the watch-glass carefully, allow the precipitate of AgCl to fall into the crucible, and remove the parting flask. Pour the water out of the crucible, remove the last portions with filter paper, and dry on a water-bath. Ignite to incipient fusion, and weigh.

Note.—The precipitate settles best in presence of an excess of $AgNO_3$.

B.—Determination of Barium.

See Fres., § 132, I, 1, and § 101, I, a. Dissolve 1 to 1.5 grm. substance in warm water ; acidulate with HCl ; dilute to about 250 c.c.; heat to boiling; when boiling hard, add dilute H_2SO_4 in slight excess ; boil some minutes and then keep warm while the precipitate settles. Test with a drop of H_2SO_4; wash with boiling water by decantation; then bring the precipitate on a No. 2 filter; wash well; dry and ignite precipitate in a platinum crucible; burn filter in a cradle as above, and add ashes to contents of crucible. See Fres., § 71, *a*.

Note. — Wash until the filtrate gives no precipitate with $AgNO_3$. When estimating barium in the presence of nitrates, chlorides, etc., these salts are sometimes carried down with the $BaSO_4$. Since it is impossible to remove these by washing with water alone, treat the precipitate with very dilute HCl, or ammonic acetate. Cf. Crookes' *Select Methods*, page 312.

C. Determination of Water (by Ignition).—

In a weighed crucible weigh out 1 to 1.5 grms. substance; heat very gently at first over a small flame, and increase the temperature very gradually; finally, heat to low redness; then cool, weigh, and repeat the operation until the weight remains constant. Caution: avoid too high a temperature, else the Cl will be expelled. When substances contain large percentages of water, as magnesic sulphate, hydrodisodic phosphate, alum, etc., begin to expel the water at 100° C. in an air-bath.

D. Calculation of Analysis.—

See Fresenius, page 795, also § 196. Make two statements, the first to determine the amount of the desired constituent in the precipitate obtained:

$$\left. \begin{array}{c} \text{Mol. Wt. of} \\ \text{precipitate} \end{array} \right\} : \left. \begin{array}{c} \text{At. Wt. of the} \\ \text{constituent} \\ \text{desired} \end{array} \right\} = \left. \begin{array}{c} \text{Actual} \\ \text{weight of} \\ \text{precipitate} \end{array} \right\} : \left. \begin{array}{c} \text{Actual} \\ \text{weight of} \\ \text{constituent.} \end{array} \right\}$$

The second statement determines the percentage of the desired constituent in the substance taken:

$$\text{Wt. of substance taken} \left.\right\} : \begin{array}{c} \text{Actual weight of} \\ \text{constituent} \end{array} \left.\right\} = 100 : \begin{array}{c} \text{Percentage of the} \\ \text{constituent.} \end{array} \left.\right\}$$

To check work, compare with theoretical percentages when possible.

Theoretical composition of crystallized barium chloride.

$$Ba = 56.15$$
$$Cl_2 = 29.09$$
$$2H_2O = 14.76$$
_____
$$100.00$$

Use of Fresenius' Tables for the calculation of analyses. Compare Table III, Fres., page 854.

Examples: $Fe_2O_3 \times 0.7 = 2Fe.$
    $BaOSO_3 \times 0.34335 = SO_3.$

Consult Table IV, Fres., pages 856, *et seq.*, also page 840.

Example: 1.2685 grms. $MgSO_4$ yielded a precipitate of $BaSO_4$, which weighed 1.2074 grms. From the table we have:

1.	0.34335
.2	0.06867
.00	0.00000
.007	0.00240
.0004	0.00013

_____   _____

1.2074 $0.41455 = SO_3$

$$\frac{0.41455}{1.2685} = 32.78 \text{ per cent. } SO_3.$$

E. Reporting Analyses.—

Analyses may be reported on blank forms printed on letter paper 8" x 10", having following headings :

HARTFORD, ——, 188 . REPORT OF ——. ANALYSIS
OF ——. DETERMINATION OF ——. GRAMMES TAKEN
——. METHOD OF ANALYSIS ——. These headings are
printed in vertical column; in one horizontal line
are placed following headings: PRECIPITATES, ACTUAL
WEIGHTS, CONSTITUENTS, CALCULATED WEIGHTS, PER-
CENTAGES, THEORETICAL PERCENTAGES; under each a blank
space is left of 2 1-2 inches. Under "precipitates" place
formulæ of precipitates obtained; under "actual weights"
place actual weights of precipitates; under "constituents"
place formulæ of constituents to be reported; under "calculated weights" place the amounts of constituents existing
in precipitates; under "percentages" place percentages of
constituents actually obtained — in short, the results of
analyses. The last column, "theoretical percentages," can
be filled only in the case of few pure chemical salts.

The words SPECIAL REMARKS are printed about two inches
from the bottom of the sheet, leaving room for remarks on
processes employed, etc.*

Notes to the Analysis of Barium Chloride.

Reactions. (1) $BaCl_2 + H_2SO_4 = BaSO_4 + 2HCl.$
(2) $BaCl_2 + 2AgNO_3 = 2AgCl + Ba(NO_3)_2.$

The chloride of silver precipitate changes color on expos-
ure to light, losing chlorine and forming Ag_2Cl; the change,
however, is only superficial, but Mulder says the loss of
weight is appreciable.

---

* See specimen blank at the end of this book.

When one part of silver is thrown down as AgCl in 1,000,000 parts of water, a slight bluish milkiness may still be seen. This cloudiness disappears on adding an excess of HCl.

Barium sulphate requires more than 400,000 parts of water for solution. The solubility is not perceptibly increased by the presence of NaCl, KClO₃ or Ba(NO₃)₂, but HCl produces a sensible increase. (Cf. Storer's *Diction-ary of Solubilities.*)

Barium sulphate thrown down in a solution containing ferric salts is often contaminated with iron. This becomes evident by the reddish color of the precipitate after ig-nition. The precipitate may be purified by washing with ammonium acetate, or by solution in conc. H_2SO_4, and re-precipitation by pouring into water. $BaSO_4$ dissolves in conc. H_2SO_4 in the ratio of 5.7 parts to 100, and in Nord-hausen sulphuric acid as 15.9 to 100.

Analysis No. 2.—MAGNESIC SULPHATE.

$$MgSO_4 + 7H_2O.$$

A.—Determination of Sulphuric Acid.

See Fres., § 132, I, 1. Dissolve 1 to 1.5 grm. of sub-stance in warm water, acidulate with HCl, dilute to about 250 c.c.; boil hard; add $BaCl_2$ carefully, avoiding a large excess; boil a few minutes; let the precipitate of $BaSO_4$ settle; wash by decantation and on the filter, and continue as in *Analysis* I, B.

B.—Determination of Magnesium.

Fres., § 104, 2.—Dissolve about 1.2 grm. of substance in 150 c.c. cold water, in a beaker; add 30 c.c. NH_4Cl,

10 c.c. NH_4HO, and a slight excess of HNa_2PO_4. (Should a precipitate form on adding NH_4HO, add NH_4Cl until it redissolves.) Stir the contents of the beaker well, avoiding touching the sides with the glass rod. Cover, and set aside for 12 hours, without warming. Filter and wash with cold water, to which one-fourth its volume of NH_4HO has been added, until the filtrate acidified with HNO_3 gives only a slight opalescence with $AgNO_3$. Dry thoroughly on the filter, ignite in a platinum crucible, gradually increasing the heat; burn the filter on a cradle until quite white before adding the ashes to the contents of the crucible. If the precipitate or ash is not white, moisten with a drop or two of conc. HNO_3, evaporate, and ignite cautiously. (See Fres., § 74, b and c.) Weigh the precipitate as $Mg_2P_2O_7$.

C. — Determination of Water.

Heat 1 to 1.5 grm. salt in a weighed platinum crucible, and proceed exactly as in *Analysis* I, C.

Notes to Analysis of Magnesic Sulphate.

On the solubility of ammonio-magnesic phosphate in water and saline solutions. Cf. Fres. page 816, paragraphs 31–35.

One part of precipitate dissolves in	15300 parts of	pure water.	
	44300 "	" ammoniated water.	
	7548 "	" strong sol. of NH_4Cl.	
	15600 "	" water containing NH_4HO and NH_4Cl.	

For a discussion of the solubility of the ammonio-magnesic phosphate, consult Gladding's letter in *Chem. News*, vol. 47, p. 71 (1883).

Reactions.—By precipitation we have :

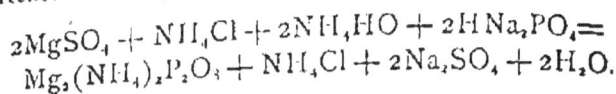

$$2MgSO_4 + NH_4Cl + 2NH_4HO + 2HNa_2PO_4 =$$
$$Mg_2(NH_4)_2P_2O_8 + NH_4Cl + 2Na_2SO_4 + 2H_2O.$$

On heating we have :

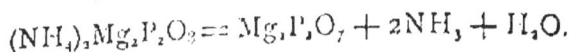

$$(NH_4)_2Mg_2P_2O_8 = Mg_2P_2O_7 + 2NH_3 + H_2O.$$

Theoretical Composition —

MgO	16.26
SO_3	32.52
7H_2O	51.22
	100.00

Analysis No. 3.—AMMONIA-IRON-ALUM.

$$Fe_2(NH_4)_2(SO_4)_4 + 24H_2O.$$

A.—Determination of Sulphuric Acid.

Dissolve 1 gr. to 1.5 grms. in water, add 5 c.c., dilute HCl to prevent ferric hydrate from precipitating with the BaSO_4, heat to boiling, add BaCl_2 and proceed exactly as in *Analysis* 2, A.

B.—Determination of Ammonium.

(Fres., § 99, *b*, 2, *β*.)

(1.) Dry the salt, if necessary, before weighing. by pressing the powder between folds of bibulous paper. Dissolve about 1.5 grms. in a little cold water in a casserole, ac'd a little dilute HCl and an excess of PtCl_4. Evaporate nearly to dryness on a water-bath scarcely heated to boiling. Add

50 to 80 c.c. alcohol to the casserole while still warm ; do not stir ; let stand several hours. The supernatant liquid should be colored by an excess of $PtCl_4$.

(2.) Place a No. 1 Swedish filter in a small funnel, wash with very dilute HCl, then with water thoroughly ; dry in the funnel, then remove the filter and place it on watch-glasses with clip ; dry in an air bath 100° C. exactly, for one hour precisely ; then close glasses and weigh the whole.

(3.) Bring the yellow crystalline precipitate on the weighed filter by means of a clean feather, wash with alcohol carefully, not too much, dry on funnel. Then transfer to clip, dry at 100° C. as before, and weigh. Dry and weigh again, repeating until constant ; calculate results. Precipitate has the composition $(NH_4)_2PtCl_6$.

[In the case of potassium determinations, wash with a mixture of alcohol and ether ; also concentrate filtrate and washings, filter from the secondary precipitate and add to the former.]

(4.) Transfer the precipitate to a weighed crucible, burn the filter and add the ashes ; ignite gradually and strongly. Weigh the Pt remaining as a check on the first determination.

[In the case of potassium, add a little oxalic acid in powder to the contents of the crucible, ignite, wash residue with water, dry on water-bath, ignite, and weigh. (See Fres., § 97, 3, β.)]

For solubility of ammonio-platinic chloride, see Fres. p. 812, paragraph 16.

C.— Determination of Iron.

I. *By Ignition.*—(Fres., § 113, 1, e.) Expose 1.0 grm. of the salt in a weighed covered platinum, or porcelain

crucible, to a moderate heat, gradually raise the temperature till all the water is expelled; then heat intensely before the blast-lamp. Weigh the residue as Fe_2O_3; heat
and weigh again. Test the residue for H_2SO_4.

II. By Precipitation. — (Fres., § 113, 1, a.) Dissolve
about 1 grm. of the salt in question in a large beaker with
about 250 to 300 c.c. of water, acidify with HCl, heat nearly
to boiling, add NH_4HO in excess; let settle after stirring;
wash hot by decantation. (N.B. — Wash out NH_4Cl completely, lest on subsequent ignition a portion of the iron
volatilize as chloride. One grm. of ferric hydrate requires
nearly one gallon of water.) Bring precipitate on filter, dry
thoroughly on funnel, ignite and weigh. Burn filter and
precipitate separately. (See Fres., § 53.)

Ammonia acts on the ferric solution in accordance with
the equation :

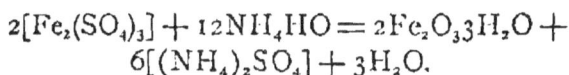

$$2[Fe_2(SO_4)_3] + 12NH_4HO = 2Fe_2O_3 3H_2O + 6[(NH_4)_2SO_4] + 3H_2O.$$

III. Determination of Iron by Marguerite's Method. — See Fres., § 112, 2, a. Compare Mohr's *Titrirmethode*,
pages 180 to 204, also Crookes' *Select Methods*, page 73.

(1.) *Standardization of the Solution of Potassium Permanganate.* — Dissolve 13 grms. $K_2Mn_2O_8$ in two litres of
distilled water, shake, let settle over night, and siphon off
into a bottle. Fill a Gay-Lussac burette with this solution
up to the zero mark.

Dissolve exactly 0.2 grm. of piano-forte wire, previously
cleaned with sand-paper, in a closed flask with conc. H_2SO_4
and sufficient water. Boil until dissolved; cool suddenly
under the faucet, but to avoid collapse of flask wait a few

moments before allowing the cold water to fall upon it.
The flask should be provided with a Krönig caoutchouc
valve. This is made by inserting a short glass tube through
a cork in the neck of the flask, and fitting to the projecting
end of the tube a piece of caoutchouc tubing about 10 cm.
long. A slit 4 to 5 cm. long is cut lengthwise in the
caoutchouc tubing, and the open end stopped with a piece
of glass rod. The valve is then complete. (Fig. 1.)

FIG. 1.　　　　　FIG. 2.

In place of the Krönig valve, another form may be used.
The projecting end of the glass tube, fitted to the cork in
the neck of the flask, is passed through another cork until
just even with its surface. Over the end of the cork and
tube a small piece of sheet caoutchouc is fastened by means
of pins, the rubber acting as the valve. (Fig. 2.) Having
effected the complete solution of the iron wire in one of

these flasks, pour the solution into a large beaker containing about 300 to 400 c.c. H_2O, placed upon a sheet of white paper; wash flask carefully, and add to beaker. Now pour the solution of $K_2Mn_2O_8$ from the burette, drop by drop, stirring continually, and continue until the pink hue first permanently colors the whole liquid. Read the burette and calculate as follows for the standard:

c.c. used: 1 c.c. $=$ grms. Fe: x, or *standard.*

Repeat the titration until two concordant results are obtained. Correction: To allow for the impurities in the iron, multiply the amount taken by 0.997.

(2.) *Reduction of the Ferric Solution.* — Dissolve 40 grms. ammonia-iron-alum in water, dilute to exactly 500 c.c.; mix well, and divide in halves.

Place a piece of amalgamated zinc and a strip of platinum foil in each reduction bottle; pour in the solutions and washings; add a little conc. H_2SO_4, and cover the bottles with watch glasses. The reduction requires six to eight hours. If the platinum foils are new, scour them with silica, rub them with KHO solution, then with HNO_3, and wash carefully. Removal of the polished and possibly greasy surface hastens the evolution of hydrogen and consequently the reduction.

Reaction:

$$Fe_2(SO_4)_3 + Zn + H_2SO_4 = 2(FeSO_4) + ZnSO_4 + H_2SO_4.$$

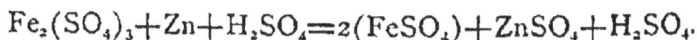

(3.) *Performance of the Analysis.* —When the reduction is complete, ascertained by testing a few drops with ammonium sulphocyanide, pour the contents of each reduction bottle into a large beaker, add H_2SO_4, and $K_2Mn_2O_8$ from the burette until a permanent pink color is obtained. (See Fres., § 112, 2, a.) The two determinations, one in each bottle, should not vary more than 0.2 per cent.

(4.) *Calculation of the Analysis.* No. of c.c. used × standard $= a$ or amount Fe.

$$\frac{a \times 100}{\text{wt. of salt taken}} = \text{per ct. of iron.}$$

IV. The standard of the solution of potassium permanganate may be determined in several ways.

(a.) *Mohr's Method.* — Weigh out 1.4 grm. ammonio-ferrous sulphate, dissolve and titrate as usual. One-seventh of its weight $=$ iron. Ammonio-ferrous sulphate $= FeSO_4 + (NH_4)_2SO_4 + 6H_2O.$

In both this and the preceding method the reaction is the same.

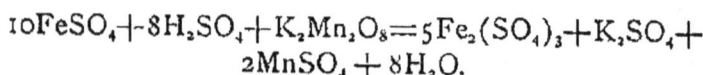

$$10FeSO_4 + 8H_2SO_4 + K_2Mn_2O_8 = 5Fe_2(SO_4)_3 + K_2SO_4 + 2MnSO_4 + 8H_2O.$$

(b.) *Hempel's Method.* — Weigh out 6.3 grms. pure, dry oxalic acid, dissolve in one litre of water, making a decinormal $(\frac{N}{10})$ solution. Dilute 50 c.c. of this solution, add 6 to 8 c.c. conc. H_2SO_4, *warm* and titrate. The reaction in this case is as follows:

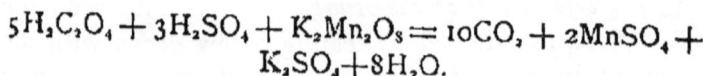

$$5H_2C_2O_4 + 3H_2SO_4 + K_2Mn_2O_8 = 10CO_2 + 2MnSO_4 + K_2SO_4 + 8H_2O.$$

D.—Determination of Water.

Water may be determined by difference.

Theoretical composition:

$$(NH_4)_2O = 5.39$$
$$Fe_2O_3 = 16.60$$
$$4SO_3 = 33.20$$
$$24H_2O = 44.81$$

$$\overline{100.00}$$

Analysis No. 4.— POTASSIUM CHLORIDE.

KCl.

Expel hydroscopic moisture carefully by heating and stirring in a porcelain dish over a Bunsen burner, before filling the weighing tube.

A.—Determination of Chlorine.

Dissolve about 0.8 grm. in warm water and proceed exactly as in *Analysis No.* 1, A.

B.—Determination of Potassium.

See Fres., § 97, 3, *a*, and Crookes' *Select Methods*, page 1.
Dissolve about 0.5 grm. in a little cold water in a casserole, and proceed exactly as in the determination of ammonia. *Analysis No.* 3, B, *paying especial attention to the sentences in brackets.*

For solubility of potassio-platinic chloride, see Fresenius' *Quant. Analysis*, p. 811, paragraph No. 8.

Theoretical composition:

K. 52.41
Cl. 47.59
                                        ─────
                                        100.00

Analysis No. 5. — HYDRODISODIC PHOSPHATE.

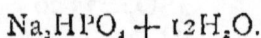

$$Na_2HPO_4 + 12H_2O.$$

A. — Determination of Sodium.

Cf. Fres., § 135, *a*, *β*.—Dissolve about 1 grm. salt in 200 c.c. water in a large beaker.

Weigh off about 0.6 grm. clean piano-forte wire, place in a flask, add conc. HCl with some HNO_3, boil hard (under a hood); when fully dissolved, continue boiling until excess of HNO_3 is removed, then dilute, and, if necessary, filter through a filter previously washed with dilute HCl.

Add this solution of pure Fe_2Cl_6 to that of the hydrodisodic phosphate, and immediately an excess of NH_4HO. Heat and let the precipitate stand some hours; wash by decantation with boiling water very thoroughly. Evaporate the filtrate with a slight excess of dilute HCl on a water-bath to dryness. Heat with care until fumes of NH_4Cl cease to come off; dissolve the residue in water; filter through a very small filter into a small weighed dish, platinum preferred. Add a few drops of dilute HCl; evaporate to dryness on a water-bath; ignite very cautiously, not too long, and weigh the NaCl. If the residue is not perfectly white and soluble in water without residue, dissolve, filter through a very small filter into another weighed dish. Evaporate and ignite again. Test residue.

B. — Determination of Phosphoric Acid.

Fres., § 134, I, b, *a*. — Dissolve about 1.2 grms. of the salt in question in cold water; add "magnesia mixture" in excess and NH_4HO; set aside for twelve hours, and then continue exactly as in *Analysis No. 2*. Consult Fres., Exp. 32, p. 817.

C. — Determination of Water.

(1) *By ignition.* — Weigh out about 0.8 gramme; place it in a weighed crucible, in an air-bath, until partially dehydrated; then heat cautiously over a Bunsen burner, ignite eventually to redness, and weigh.

(2) *By direct weight.* — Weigh out about 0.7 gramme substance, and introduce it into the weighed ignition bulb by means of a very narrow piece of folded paper. Weigh $CaCl_2$ tube, and arrange apparatus, as shown in Fig. 25, page 61, of Fres. *Quant. Analysis* (American edition, 1881), substituting aspirator for gasometer if more convenient. Heat cautiously, aspirating continually, and raise the temperature to a low red heat for three minutes. In driving the water into the $CaCl_2$ tube be careful not to burn the cork. Aspirate while cooling, not too rapidly. Weigh $CaCl_2$ tube after cooling and the ignition bulb as a *check.* Consult Fres., § 36.

Theoretical Composition:

When water is determined by heating to redness, the calculation must be based on two molecules of the salt.

$$2Na_2O = 17.32$$
$$P_2O_5 = 19.83$$
$$25H_2O = 62.85$$
$$\overline{100.00}$$

*Analysis No. 6.—*SILVER COIN. SCHEME. $Au(?) + Ag + Pb + Cu$.

Clean the coin by friction with wood ashes and weigh it. Dissolve in HNO_3 in a covered casserole, expel excess of acid by evaporation to small bulk on a water-bath, add water, filter, and wash thoroughly.

Residue a.

Dry on filter, ignite, and weigh as $Au(?) + Ag_2S$.

Filtrate a.

Heat filtrate together with wash water to boiling; add dilute HCl in excess, agitate well, let settle till perfectly clear, filter and wash.

Precipitate b.

Dry, ignite separately from filter, and treat exactly as in *Analysis No.* 1, A. Weigh as AgCl, and calculate Ag. Cf. Fres., § 115, 1, a, β, also Fres., § 82, b.

Filtrate b.

Add 3 c.c. dilute H_2SO_4 to filtrate + washings; evaporate nearly to dryness on a water-bath; filter through a No. 1 filter, wash with as little water as possible, and yet completely.

Precipitate c.

Dry, ignite in porcelain crucible, burning the filter on cover. Weigh as $PbSO_4$. Cf. Fres., § 116, 3, a, β, also Fres., § 83, d. See also Crookes' *Select Methods*, p. 209.

Filtrate c.

Heat to boiling in a casserole, concentrate if necessary, add pure KHO solution in excess, boil some minutes, or until the bluish precipitate becomes quite black, wash hot by decantation, bring on filter, cleanse casserole with a rubber-tipped glass rod, or with the tip of your little finger.

Precipitate d.

Dry, ignite, and weigh as CuO. If some CuO is reduced by filter paper, add a drop of HNO_3, evaporate and ignite; this may, however, occasion loss. Cf. Fres., § 119, 1, a, α.

Filtrate d.

Test with H_2S, passing the gas through solution some minutes; if precipitate forms let settle, collect on filter, dry and ignite. Weigh as CuO. (By roasting CuS is largely oxidized.) See Fres., § 55, b, and § 119, 3, a.

Analysis No. 7.—Dolomite. Skeleton Scheme.

$$CaCO_3 + MgCO_3.$$

Dissolve, evaporate, and filter. (*See Note 1.*)

Residue a.
SiO_2
(See *Note 2.*)

Filtrate a.
Throw down iron and alumina. (*Note 3.*)

Precipitate b.
$Fe_2O_3 + Al_2O_3$.

Dissolve, reprecipitate and add filtrate to *Filtrate b.*
(*Note 4.*)

For determination of CO_2 see *Note* 8.

Filtrate b.
Throw down calcium. (*Note 5.*)

Precipitate c.
CaC_2O_4.
Note 6.

Filtrate c.
Throw down magnesium.
(*Note 7.*)

For calculation see *Note* 9.

NOTES TO FOREGOING SCHEME.

Note 1.—Take 1.5 to 2.0 grammes finely powdered mineral, dissolve in dilute HCl in a casserole; heat, add a little HNO_3 to oxidize iron and sulphides; evaporate to dryness on a water-bath; moisten with HCl, add water, digest, and filter from the SiO_2 + Silicates. Dry on funnel, ignite, and weigh. See note 2.

Note 2.—If it is desirable to determine the SiO_2 in the silicates present, "*Residue a*" must be treated as follows : Dry and ignite (with filter), mix in a platinum crucible with about six parts of Na_2CO_3 (anhydrous), and fuse at a red heat. Cool, remove the fused mass with boiling water, add an excess of HCl, evaporate to dryness on a water-bath, heat in an air-bath until the HCl is *completely* expelled ; again moisten with HCl, dissolve in water, and filter from the residue. The residue which is now pure hydrated SiO_2, is dried, ignited, and weighed. The filtrate must be added to "*Filtrate a.*" Examine Fres., § 140, II, *b*, *a*, and § 93, 9.

Note 3.—Heat the filtrate from "*Residue a*," add NH_4Cl, and NH_4HO in slight excess. (The NH_4Cl may be omitted if the "*Filtrate a*" is very acid.) Heat until excess of NH_4HO is expelled, filter quickly, and wash hot. *See Fres.*, § 113, 1, *a*, *and* § 105, 1, *a*.

Note 4.—"*Precipitate b*" is partly washed, and then, while moist, dissolved in a little warm dilute HCl on the filter, the solution is reprecipitated by NH_4HO and the precipitate brought on the same filter, washed thoroughly, dried, and ignited. Weigh as $Fe_2O_3 + Al_2O_3$. The second filtrate is added to "*Filtrate b.*"

Note 5.—Concentrate "*Filtrate b*," add some NH_4Cl unless present already, add $(NH_4)_2 C_2O_4$ in considerable excess, and some NH_4HO. Let stand 12 hours in a warm place. Wash partially and filter. *See Fres.*, § 154, 6, *a* ; also § 103, 2, *b*, *a*.

Note 6.—Dissolve the partially washed "*Precipitate c*" in HCl, reprecipitate with NH_4HO and a little $(NH_4)_2C_2O_4$. Filter and wash hot, add filtrate and washings to "*Filtrate c.*" Dry precipitate on funnel, transfer to crucible, burn filter, add ashes, add a few drops of conc. H_2SO_4 to contents of crucible, ignite cautiously to low redness, and weigh as $CaSO_4$. Compare Fres., § 103, 2, *b*, *a*.

Note 7.—If care has been taken to avoid undue excess of NH_4Cl in the preceding steps, the magnesium may be thrown down in "*Filtrate c*" immediately. Otherwise the NH_4Cl must be expelled as follows : Concentrate the liquid, add 3 grms. of HNO_3 for every grm. of NH_4Cl supposed to be in the solution, warm gently (60° C.) and eventually heat to boiling.

Concentrate "*Filtrate c*" add NH_4HO and Na_2HPO_4 and proceed as in *Analysis* 2. B. See Fres., § 104, 2, and § 74.

Notes on the Decomposition of NH_4Cl *by* HNO_3 *in solution.* *Comptes Rendus*, October 13, 1851 (Maumené). *J. Lawrence Smith* in *American Chemist*, Vol. III, p. 201. Also *Am. Jour. Sci.* (2), Vol. 15, note, page 240, which is as follows : " The character of the decomposition which takes place is somewhat curious and unexpected : it was first supposed that equal volumes of Cl, N_2O, and N were given off, but it is shown that nearly all the NH_4HO, with its equivalent of HNO_3, is converted into N_2O, the liberated HCl mixing with the excess of HNO_3. A little of the $NH_4Cl+HNO_3$ does not undergo the decomposition first supposed, and in this way only can the small amounts of N and Cl be accounted for." " Some nitrous or hyper-nitrous acid forms during the whole process if conc. HNO_3 is used little or none if dilute HNO_3."

The action of NH_4NO_3 on NH_4Cl is theoretically as follows :

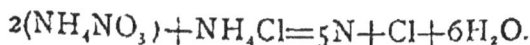

$$2(NH_4NO_3)+NH_4Cl=5N+Cl+6H_2O.$$

The following are possible reactions:

$$8NH_4Cl+10HNO_3=9N_2O+8Cl+21H_2O,$$
$$2HNO_3+2NH_4Cl=N_2O+2Cl+2N+5H_2O,$$
$$HNO_3+NH_4Cl=HCl+N_2O+2H_2O,$$
and
$$2HNO_3+NH_4Cl=N_2O+Cl+NO_2+3H_2O,$$
and
$$HCl+3HNO_3=NO+Cl+NOCl_2+NO_2+4Cl+5H_2O.$$

Note 8. *Determination of* CO_2.—*I. By loss.* Fres., § 139, II., *d, bb,* and *cc.*

Weigh out 1.0 to 2.0 grms., place in the Geissler apparatus, fill the proper portions of the apparatus with HCl (dil.) and with H_2SO_4 (conc.) respectively. Weigh apparatus. Cautiously let the HCl flow on the mineral, warm gently, heating at the last till the solution begins to boil. Cool apparatus and weigh. For details consult *Fresenius*, as above. Do not hurry this process too much.

Fig. 3

II.—By direct weight. Consult Fres., § 139, II., c.

Arrange apparatus as in Fig. 4. Suspend tubes by wire loops on nails.

a contains soda-lime.

c is a flask of about 200 c.c. capacity.

d contains conc. H_2SO_4.

e contains pieces of pumice-stone saturated with conc. H_2SO_4; avoid much liquid in the bend.

f contains pumice-stone saturated with anhydrous $CuSO_4$,

N.B.—Make a strong hot solution of $CuSO_4 + 5H_2O$, add pieces of pumice-stone, boil hard, evaporate to dryness and ignite well. The product should be nearly white.

g contains in outer tube, soda-lime; in inner tube, (h) pumice-stone saturated with H_2SO_4; weigh these together both before the absorption and after.

Fig. 4.

Place 1.0 to 1.5 grms. mineral in c, weigh g and h, and connect apparatus; a is not attached at first. Pour a little water through the funnel tube into c, then add gradually HCl, diluted one-half with water. Attach a, and aspirate gently. Heat cautiously to incipient ebullition; maintain this a few moments, and let cool while the aspiration continues. Weigh — increase of weight gives CO_2.

Note 9. *Calculation.*—Normal dolomite contains :

30.4 per cent. CaO.
47.8   "   CO .
21.8   "   MgO.
_____
100.0

Having estimated these constituents, calculate the

amounts of $CaCO_3$ and $MgCO_3$, and report under "Special Remarks," thus :

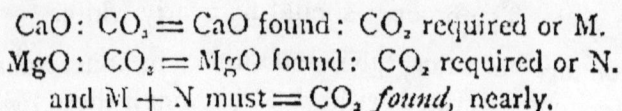

$CaO : CO_2 = CaO$ found : CO_2 required or M.

$MgO : CO_2 = MgO$ found : CO_2 required or N.

and M + N must $= CO_2$ *found,* nearly.

Analysis No. 8 — BRONZE.

To be determined, *Sn, Pb, Cu, Zn.*

A.—Determination of Tin.

Dissolve about 0.6 grm. bronze filings, carefully freed from accidental impurities, in moderately dilute HNO_3, in a flask in the neck of which is placed a small glass funnel. After complete solution (except the SnO_2), transfer contents to a porcelain dish, evaporate to dryness, moisten with HNO_3, add H_2O, and filter from the SnO_2. Dry this residue, ignite in porcelain, and weigh. Fres., § 126, I., *a,* and § 91.

B.—Determination of Lead.

To filtrate from A add dilute H_2SO_4, evaporate until fumes of H_2SO_4 appear, or the residue is nearly dry, let the dish cool, then add water, and filter from the $PbSO_4$. See Fres., § 163, 2, and § 116, 3, a, *β.* Dry, ignite, and weigh precipitate. See Fres., § 83, *d.*

C.—Determination of Copper.

The filtrate from B. should not measure more than 100 c.c. Place the solution in a large platinum dish, arrange the Bunsen cells of a galvanic battery, connect the zinc element with the platinum dish, and the carbon element with a small piece of platinum foil which is immersed in the liquid. Let the battery run four or five hours. Take out a drop of the solution with a pipette, place on a watch glass and test for Cu with H_2S. Pour out the solution when the precipitation is completed, and wash thrice with small quantities of water. Then wash the copper film with alcohol twice, dry in the hand, over a Bunsen burner, at a very gentle heat, and weigh quickly.

N.B.— It is advisable to test solution for Cu before proceeding further.

D.— Determination of Zinc.

Heat the filtrate and washings from C to boiling, add excess of Na_2CO_3, boil a few minutes, wash by decantation hot, then on filter, Dry, ignite, and weigh as ZnO. Fres., § 108, 1, a, and § 77.

Analysis No. 9.— COAL. (PROXIMATE ANALYSIS.)

To be determined, Moisture, Volatile and Combustible Matter, Fixed Carbon, Sulphur, and Ash.

A.— Determination of Moisture.

Pulverize the coal very finely, heat one to two grms. in a half ounce platinum crucible for fifteen minutes at 115° C. in an air-bath, cool and weigh. Repeat this desiccation in the air-bath, weighing at intervals of ten minutes, until the weight is constant or begins to rise. Loss of weight gives moisture. In reporting, give exact temperature at which it was determined. N.B. — The increase in weight is due to oxidation of the coal; it generally begins after heating thirty to ninety minutes in the air-bath. *Anthracite coal* may be heated an hour or more. See *Chem. News*, Am. Repr., Vol. V., p. 80.

B.—Determination of Volatile Combustible Matter.

Heat the same crucible with contents, closely covered, to bright redness over a Bunsen burner, exactly three and one-half minutes, and then without allowing the crucible to cool, heat strongly before the blast-lamp, exactly three and one-half minutes more. Cool and weigh. The loss gives the volatile and combustible matter, and includes half the S in the FeS_2. See F *below*.

C.—Determination of Fixed Carbon.

Heat crucible and contents, uncovered, over Bunsen burner, until all carbon is burned off and the weight is constant. This takes from one to four hours or more. Loss in weight = fixed carbon, including half the S.

D.—Determination of the Ash.

The difference between the weight last obtained and that of the crucible gives the weight of the ash. Note color of the ash.

E.—Determination of Sulphur.

Secure a sample of anhydrous Na_2CO_3, shown to be absolutely free from S by the silver test.

Weigh out about two grms. coal in fine powder, mix with about ten grms. $NaNO_3$ and ten grms. Na_2CO_3 on glazed paper. The sodium salts need not be weighed accurately ; KNO_3 may be used in place of $NaNO_3$. Deflagrate in a covered two-ounce platinum crucible, heating over a Bunsen burner; add the mixed coal and sodium carbonate little by little, replacing the cover of the crucible quickly each time. Do not expect to effect a perfect fusion. Place the crucible and contents in a casserole, add water, and digest until the mass is disintegrated, and the crucible can be removed. Add cautiously an excess of HCl, heat to boiling, and throw down the H_2SO_4 with $BaCl_2$ as usual. If flocks of SiO_2 remain insoluble in HCl, evaporate to dryness on water-bath, heat until HCl is expelled, add water, filter, and proceed as above. If the $BaSO_4$ is reddish after ignition, wash with solution $NH_4C_2H_3O_2$ and then with pure water, dry, ignite, and weigh again. The $BaSO_4$ may also be purified by solution in conc. H_2SO_4 and reprecipitation with water.

Second Method for Determining Sulphur. — Put two to five grms. powdered coal in a flask holding a litre; add 100 c.c. HNO_3 and five grms. powdered $KClO_3$, heat to boiling, adding more reagents as needed; continue until all the carbon is oxidised. Transfer to a dish, evaporate to dryness, add HCl and water, throw down H_2SO_4 with $BaCl_2$, and proceed as usual. Consult Hayes's article in *Am. Chem.*, Feb., 1875, also Wittstein's article in *Am. Chem.*, April, 1876.

F. — Calculations.

Theoretically we should deduct half S from the volatile combustible matter (because iron pyrites loses one-half its sulphur at a red heat), one-eighth S from the fixed carbon, and three-eighths from the ash. ($2FeS$ become Fe_2O_3, or $8 \times 4 = 32$ reduces to $8 \times 3 = 24$.)

Practically half the amount of sulphur is deducted from the volatile combustible, and half from the fixed carbon; reports should be made out accordingly.

G. — Estimation of Carbon and Hydrogen.

Ignite one grm. of coal with $PbCrO_4$ in a hard glass tube 0.25 metre long. Pass the H_2O, CO_2 and H_2SO_4 formed through two U-tubes, one containing ignited $CaCl_2$, and the other a solution of $Pb(NO_3)_2$, and through a potash bulb. The increase in weight of the first U-tube gives the H_2O, and that of the potash bulb the CO_2.

Calculation of Calorific Power. — One part of carbon in burning yields 8,080 calorific units, and one part of hydrogen in burning 34,460 calorific units. Hence to calculate total calorific units in a coal, multiply the percentage of C by 8,080 and divide by 100; also multiply the percentage of H by 34,460 and divide by 100. Add the quotients.

(A calorific unit is the amount of heat necessary to raise one grm. of water from 0° to 1° C.) See *Chem. News,* XXXIV, p. 233. 1876.

Analysis No. 10. — COPPER PYRITES.

Determination of Copper.

Pulverize *very* finely. Weigh out exactly 2 grms., place it in a flask of about 300 c.c. capacity and covered with a small funnel, the stem of which is slipped into the neck of the flask. Add 20 c.c. conc. HNO_3, 5 c.c. conc. HCl, mixing these in flask under the hood. Digest some minutes, then add cautiously 20 c.c. conc. H_2SO_4 and boil hard until fumes of H_2SO_4 appear abundantly. Cool, add water with caution, dilute not too largely, filter from residue (SiO_2, $CaSO_4$, etc.), and wash. Test residue for copper before the blow-pipe. Dilute filtrate to 200 c.c. exactly, mix well by pouring into a dry beaker and back again three or four times; divide in halves by taking out 100 c.c. with a pipette and place in a platinum dish previously weighed. (N.B.—. Volumetric apparatus as sold is rarely reliable, therefore test pipette and flask before measuring as above.) Arrange two cells of a Bunsen battery, placing the "battery acid" (one part of H_2SO_4 diluted with 8 to 10 of water) in the outer cell and "battery fluid" ($K_2Cr_2O_7 + H_2SO_4 + H_2O$) in the inner. Connect the zinc ($+$) pole with the platinum dish, and the carbon ($-$) pole with a piece of platinum foil which is immersed in the liquid. Cover the platinum dish with two pieces of glass plate, one each side of the platinum foil, to prevent loss by spattering. Or use the cone or spiral described in *Chem. News*, XIX, p. 222 (1869). See also Crookes *Select Methods*, pages 187–200.

It is best not to let the battery run all night; prepare the solutions on one day and start the battery the next morning. Four hours or more usually suffice for complete precipitation.

Test a few drops of the solution with H_2S.

When precipitation is complete, pour off liquid, wash copper with distilled water three or four times (work rapidly), then with strongest alcohol twice; drain the alcohol off, dry the copper at a very low heat, holding the platinum dish in the hand over a small flame, which must not touch the dish, and weigh immediately. Next treat the remaining 100 c.c. solution likewise; the two determinations should agree to about 0.2 per cent.

Analyses No. 11 *and No.* 12.

Introductory Notes on Volumetric Analysis.

Definition. "Volumetric Analysis is a form of quantitative analysis in which we seek to estimate the amount of a substance from the determinate action of reagents in

solutions of known strength, the amount of the reacting
substance being calculated from the volume of the liquid
used." The first principles and method of procedure have
been foreshadowed in *Analysis No.* 3, III., Determination
of Iron by Marguerite's method. For explanation of general volumetric methods, see Fres. § 54, and consult *Sutton's Handbook of Volumetric Analysis*, also Mohr's *Lehrbuch der chemisch-analytischen Titrirmethode*.

Principles. When volumetric analysis first came into
use, the standard solutions were so prepared as to give
results in percentages ; thus in Alkalimetry, one standard
solution of acid was used for potash, another for soda, etc.
The modern system is based on the fact that acids and
alkalies (as well as other reagents) neutralize each other in
the proportion of their molecular weights, or of simple
multiples of the same ; consequently standard solutions are
so prepared that *one litre* contains one-half or the whole
of the molecular weight of the reagent *weighed in grms.*
For example, the molecular weight of HCl being 36.5 and
that of KHO 56.1, 36.5 grms. of HCl exactly neutralize
56.1 grms. of KHO, and if these respective amounts be
dissolved in one litre of water, the whole of one solution
will not only neutralize the whole of the other, but any
aliquot part of one will exactly neutralize a similar aliquot
part of the other. And by using graduated vessels, (burettes,) the amount of reagent used is determined by the
volume of the solution. (Before employing burettes,
pipettes, and graduated flasks, care should be taken to test
the accuracy of the graduation.)

Standard Solutions. Solutions containing the molecular weight of the reagent expressed *in grms. per litre* are
called *normal* solutions ; in the case of di-basic acids
(H_2SO_4, $H_2C_2O_4$ etc.) and of "di-acid" alkalies (Na_2CO_3)

one-half the molecular weight of each is taken, making *half normal* solutions.

The standard solutions of the following reagents are made with the quantities indicated :

Oxalic acid	$H_2C_2O_4 + 2$ aq.	63 grms.	per litre
Sulphuric acid	H_2SO_4	49 "	"
Hydrochloric acid	HCl	36.5 "	"
Sodium carbonate	Na_2CO_3	53 "	"
Potassium hydrate	KHO	56.1 "	"
Ammonia	NH_3	17 "	"

The point of neutralization or *end reaction* is determined by adding to the solutions some organic coloring-matter which changes in hue under the influence of an alkali or an acid. The " indicators " commonly used are litmus solution and cochineal solution.

ALKALIMETRY.

(Cf. Sutton's Handbook.)

1. *Preparation of Litmus Solution.*—Digest 5 to 6 grms. litmus with about 200 c. c. water for half an hour or more; decant the clear liquid or filter ; add very dilute HNO_3 drop by drop, until the color is changed to violet. If properly neutralized less than one-tenth c.c. of standard acid should distinctly redden one c.c. litmus in 100 c.c. of water.

2. *Sulphuric Acid.*—Mix about 60 grms. conc. C.P.

H_2SO_4 of sp. gr. 1.840 with three or four times its volume of distilled water ; cool and dilute to one litre. The exact standard of this solution is determined by testing with sodium carbonate, as below.

3. *Sodium Carbonate Solution.* — Weigh off about 12 grms. anhydrous C.P. Na_2CO_3 ; heat in a porcelain dish to low redness, stirring until moisture is expelled ; place in a desiccator to cool. Weigh out accurately 10.6 grms. of this, and dissolve in distilled water. Dilute to exactly 200 c.c. This gives a half normal solution, each c.c. of which contains 0.053 grm. of sodium carbonate, as shown by this simple calculation :

$$Na_2 = 46$$
$$C = 12$$
$$O_3 = 48$$

Mol. wt. of Na_2CO_3 106

One-half the mol. wt. $= 53$

200 c.c. : 1 c.c. $= 10.6$ grms. : 0.053 grms.

This solution serves to standardize the sulphuric acid.

Standardizing the Sulphuric Acid. — Take of the Na_2CO_3 solution, 20, 30, or 40 c.c., accurately measured, place in a wide-mouthed flask of about 300 c.c. capacity ; add litmus solution, and run in H_2SO_4 solution from a burette until a wine-red color is obtained ; boil hard to expel CO_2, and add more acid until the color is permanent. Read off the c.c. used. Repeat the process. Suppose 30 c.c. Na_2CO_3 solution required 25 c.c. H_2SO_4 solution. Then 5 c.c. (30 — 25) water must be added to every 25 c.c. of the acid solution to make it normal. Measure, therefore, the H_2SO_4 solution carefully and add the necessary amount of water. Suppose the H_2SO_4 solution measures 900 c.c., since $900 = 25 \times 36$, then 36×5, or 180 c.c. water must be added. Add the water, mix well, and again determine

the standard : one c.c. of the Na_2CO_3 solution should exactly neutralize one c.c. of the H_2SO_4 solution.   In case of
difficulties the exact standard of the acid should be determined gravimetrically by precipitating 10 or 20 c.c. with
$BaCl_2$, and calculating from the $BaSO_4$ obtained the
amount of H_2SO_4 in one c.c.

Carminic acid being stronger than carbonic acid, a solution of cochineal is sometimes substituted for litmus, in
which case boiling may be dispensed with.   The dyestuff
tropacoline has recently been proposed as an indicator in
alkalimetry.   Cf. *Ber. d. chem. Ges.* XI, 460 (1878).

Deci-normal Solution of Acid.— Call the above normal
solution " No. 1 ;" take 100 c.c. of No. 1, put into a litre
flask, and dilute to one litre.   Call this *deci-normal solution*
" No. 2."

A.—Valuation of Soda Ash.
(Determination of Na_2CO_3.)

Place about 12 grms. powdered sample in a platinum
crucible or porcelain dish ; heat moderately for some minutes over a Bunsen burner, until all moisture is expelled ;
cool, weigh out exactly 10 grm.; dissolve in water; dilute
to one-half litre and mix well.   Take out 50 c.c. solution
(which contains one grm. soda ash), and determine the
amount of normal acid needed to neutralize, adding litmus
as before, and boiling to expel CO_2.

Suppose 50 c.c. solution soda ash required 15 c.c. stand-
ard acid, then $\frac{0.053 \times 15 \times 100}{1 \text{ grm.}} = 79.5$ per cent. Na_2CO_3.
See Fres., § 195, p. 692.   These results are only
approximative and preliminary, and the operation must
be repeated, finishing with the deci-normal solution No.
2, as below.   Take another 50 c.c. of soda ash solution;
run from a burette 12 c.c. of solution " No. 1," and then

finish with solution " No 2." Of course, in calculating, 10 c c. of No. 2 equals one c.c. of " No. 1."

B.—Valuation of Pearl Ash.

Proceed as before ; weigh quickly the salt cooled in a desiccator, for it is very hydroscopic. In calculating, use the factor 0.0691.

The Residual Method of Titration.—This method has great advantages over the foregoing method, especially when carbonates are in question ; the sharpness of the end reaction being much increased by the absence of CO_2. The process is as follows : Super-saturate the soda ash solution with normal acid in excess ; then add normal potassic hydrate (and decinormal also) until the neutral point is reached. (The normal KHO is mentioned in the next paragraph.) Since one c.c. acid = one c.c. alkali, substract the number of c.c. of standard alkali from the number c.c. of standard acid added in the first place, and then calculate as usual.

ACIDIMETRY.

Generalities.—The value of strong acids, especially HCl, HNO_3, H_2SO_4, is frequently deduced from the Specific Gravity as determined by the hydrometer. See tables in Fres., pp. 690, 691, showing percentages of acids in solutions of different densities. When titration is desirable, standard KHO solution is used, and in accordance with the principles already stated.

Preparation of Standard Alkali.—Take about 60 grms. KHO, dissolve in 1 litre of water, add $Ca(HO)_2$ to throw down carbonates, boil, let settle, and syphon off. Determine the exact standard of this with normal and decinormal acid.

A.—Valuation of HCl.

Take 5 to 50 c.c. acid, according to strength, dilute to a definite volume, take an aliquot part, add litmus and run in the standard KHO as described.

In calculating multiply the number of c.c. of KHO added by .0365 × 100, and divide this product by the number of c.c. of acid taken × Specific Gravity of the solution as determined by the hydrometer.

Example.—Took 10 c.c. HCl solution, having a Specific Gravity = 1.025 ; since 1 c.c. of water weighs 1 grm., the weight of acid taken = 10.25 grms.

The acid solution required 8 c.c. KHO, whence

$$\frac{8 \times .0365 \times 100}{10.25} = 2.84 \text{ per cent.}$$

B.—Analysis of Vinegar.

A. Determine the acetic acid by titration, using cochineal solution, or with methylaniline violet, as in the "Witz method" (*Am. Chem.*, Vol. VI, page 12), or use *Mohr's* method, as follows:

Add to a known quantity of acid a weighed quantity (in excess) of pure precipitated dry $CaCO_3$. After decomposition is nearly complete in the cold, boil to expel CO_2, filter, and wash the excess of $CaCO_3$ in hot water. Dissolve the $CaCO_3$ in excess of normal HCl, and determine the HCl remaining by means of normal KHO, or NaHO and litmus solution. The results with dark colored vinegars are good.

B. Determine water by drying at 100° C. to constant weight, and allow for alcohol and acetic acid.

C. Determine alcohol by neutralizing about 300 c.c. vinegar with $CaCO_3$ and distilling off some measured amount, say 150 c.c. Then determine specific gravity by weighing, and from this calculate the per cent. of alcohol.

D. Determine the grape sugar. (See *Analysis No.* 33.)

Analysis No. 13.—CHLORIMETRY.

Constitution of Bleaching Powder.

Bleaching powder is formed thus :

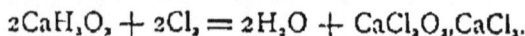

$$2CaH_2O_2 + 2Cl_2 = 2H_2O + CaCl_2O_2,CaCl_2.$$

The composition of bleaching powder is variously given. The following are some of the formulæ.

"Quelques Chimistes," $CaCl_2 + H_2O_2$.
Watts, $CaClO + CaCl$, $Ca_2O + 2H_2O$.
Bloxam, $CaO\ Cl_2O + CaCl$, $2CaO + 4H_2O$.
Roscoe, $CaCl_2O_2$.
Muspratt, $CaO\ Cl$, $2H_2O$.
Fownes, $CaCl_2 + CaCl_2O_4$.
Calvert, $2CaCl_2 + CaCl_2O_2$.
Thorpe, $Ca_2H_6O_6Cl_4 = CaCl_2O_2 + CaH_2O_2 + CaCl_2 + 2H_2O$
Kolb, $(2CaO,Cl,H_2O)$, CaH_2O_2.
Rose, $(CaCl_2, Ca_2O_2)\ CaO,Cl_2 + 4H_2O$.

Stahlschmidt's theory of its formation : *Bericht D. Chem. Ges.*, 1875 :

$$3CaH_2O_2 + 4Cl = CaCl_2 + CaCl_2O_2 + CaH_2O_2 + 2H_2O.$$

See paper on Constitution of Bleaching Powder, by Dr. Lunge in *American Chemist*, Vol. V, page 454.

When allowed to stand in contact with air and light, it decomposes, $CaCl_2$ increasing, and the $CaClO$ decreasing. Dry chloride of lime, at 50° C., decomposes thus :— (THORPE.)

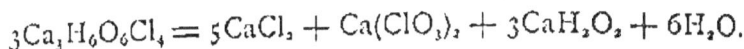

$$3Ca_2H_6O_6Cl_4 = 5CaCl_2 + Ca(ClO_3)_2 + 3CaH_2O_2 + 6H_2O.$$

By the action of water chloride of lime decomposes thus :

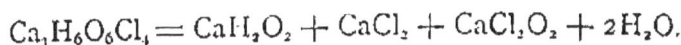

$$Ca_2H_6O_6Cl_4 = CaH_2O_2 + CaCl_2 + CaCl_2O_2 + 2H_2O.$$

The value of the commercial article depends wholly upon the amount of "available chlorine," viz.: the Cl in the hypochlorite, which is thus constantly varying.

The strongest contains 38.5 per cent. available chlorine. One or two per cent. of this is present as calcium *chlorate*, which is without bleaching power.

Action of Acids on Bleaching Powder. — Action of hydrochloric acid :

$$(CaCl_2O_2 + CaCl_2) + 2HCl = 2CaCl_2 + 2(HClO).$$

Action of dilute sulphuric acid :

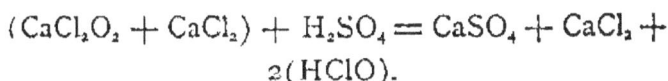

$$(CaCl_2O_2 + CaCl_2) + H_2SO_4 = CaSO_4 + CaCl_2 + 2(HClO).$$

Further action of concentrated sulphuric acid :

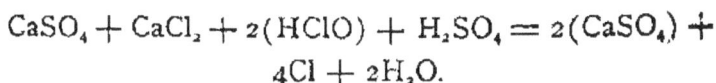

$$CaSO_4 + CaCl_2 + 2(HClO) + H_2SO_4 = 2(CaSO_4) + 4Cl + 2H_2O.$$

Valuation of Chloride of Lime.

Penot's Method. From Fresenius' *Quant. Analysis,* § 200. Based on the conversion of an alkaline arsenite, into an arseniate by a solution of chloride of lime.

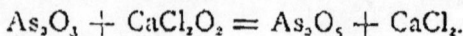

$$As_2O_3 + CaCl_2O_2 = As_2O_5 + CaCl_2.$$

The end reaction is determined by KI and starch, undecomposed hypochlorite turning this mixture blue.

(a.) *Preparation of KI Starch Paper.* — Boil three grms. starch, in 250 c.c. water, add one grm. KI, one grm. Na_2CO_3 + aq.; dilute to 500 c.c. Moisten paper with this solution and dry.

(b.) *Preparation of Solution of As_2O_3.* — Dissolve exactly 4.95 grms. pure sublimed As_2O_3 with 25 grms. Na_2CO_3 + aq. (free from S) in 200 c.c. water. Boil until dissolved and dilute to one litre. Make a $\frac{N}{10}$ solution. Since it is difficult to weigh out exactly this amount, take any number and dilute proportionately. If 5.013 grms., then

$$4.95 : 1000 = 5.013 \text{ grms.} : 1012.7.$$

Add then 12.7 c.c. to the litre. One c.c. of this solution = 0.00355 Cl.

(c.) *Process of the Determination.* — Mix sample well; weigh out 10 grms., rub in mortar with 50 or 60 c.c. water; settle; decant turbid liquid into a litre flask. Repeat. Fill up to mark, and mix.

Fill a burette, take 50 c.c., run it into a beaker, add the standard As_2O_3 solution, stirring until a drop of the solution no longer gives a blue mark on the KI starch paper. Repeat on fresh amount. *Caution:* Shake, and draw off turbid liquid.

(d.) Calculation.

$$\frac{c\,c.\ As_2O_3\ solution\ used \times 0.00355 \times 100}{Amount\ taken} = per\ cent.\ Cl.$$

[French chlorimetrical degrees represent the number of litres of Cl at 0.°C. and 760 m.m., which one kil. of sample should yield. Now one litre of Cl weighs 3.177 grms.; hence 31.77 per cent. = 100 degrees. See foot-note on p. 700 of Fres. *Quant. Anal.*]

The amount of *calcium chloride* present may be determined by first estimating the hypochlorite as above, and then adding to the second portion of 50 c.c. a slight excess of NH_4HO and warming.

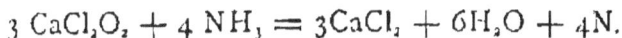

$$3\ CaCl_2O_2 + 4\ NH_3 = 3CaCl_2 + 6H_2O + 4N.$$

Neutralize the solution with HNO_3 and determine the Cl by $AgNO_3$.

The amount of *chlorate* may be determined by heating a third portion with ammonia, then acidulation with pure H_2SO_4 and digesting with Zn.

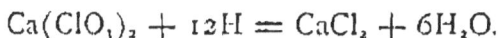

$$Ca(ClO_3)_2 + 12H = CaCl_2 + 6H_2O.$$

Again determine the Cl by $AgNO_3$, and the increased amount over the second determination gives the Cl existing as *chlorate.* (THORPE.)

Analysis No. 14.—TYPE METAL. To be determined Pb, Sb, Sn (Zn and Fe?).

Dissolve about 1 grm. clippings in moderately conc. HNO₃, adding enough H₂C₄H₄O₆ to hold up the antimony, and heating gently. Digest and add acid until all but the SnO₂ is dissolved. See Fres., § 164, 14. *b.* Expel excess of HNO₃ by concentration of liquid, but not to dryness. Filter and wash. *See Note* 1.

Residue a.

SnO₂ (+ Sb?) Dry, ignite, and weigh.
Test for lead.

See Fres., § 126, 1, *a*, and § 91, *a.*
See also *Note* 2, below.

Solution a.

Sb, Pb, Zn (Fe). Add H₂SO₄, and evaporate to small bulk, add alcohol and let stand 12 hours. See Fres.; § 116, 3, *a. a.*, and § 83, *d.* Filter and wash with water containing a little H₂SO₄, and then changing recipient of filtrate, wash thoroughly with alcohol. Be careful to expel all H₂SO₄ from filter.

Precipitate b.

PbSO₄.
Dry, ignite in porcelain, and weigh.
See Fres., § 83, *d.*

Filtrate b.

Sb, Pb, Zn (Fe). Saturate with H₂S gas, warming solution with a current of steam. Filter and wash. See Fres., § 164, A, 1.

Precipitate c.

Sb₂S₃ + PbS. Digest with yellow NH₄HS, filter, repeat digestion, filter and wash. See Fres., § 164, 14, *b.*

Solution c.

Zn and Fe.
(To determine these, throw down Fe as basic acetate, and Zn as carbonate. Fres., § 108, 1, *a*, and § 77, *a.*)

Solution d.

Add excess of HCl and wash.

Residue d.

PbS, oxidize with HNO₃, dry, ignite, and weigh as PbSO₄.

Solution c.

NH₄Cl.
Reject.

Precipitate e.

Treat precipitate of Sb₂S₃ + S on filter (to remove S) by washing with CS₂, transfer to a weighed porcelain crucible, add more fuming HNO₃, heat, add more acid, evaporate to dryness, ignite, and weigh as Sb₂O₄. If dark colored add more HNO₃, heat, ignite, and weigh again. Fres., § 135, 2, *b. c.* *See Note* 1.

Note 1. — Some of the tin may go into solution as nitrate of tin, if the nitric acid be dilute, and thus appear in *Precipitate c* mixed with the sulphide of antimony ; in this case they should be separated by F. W. Clarke's method, which is based on the solubility of the sulphide of tin in oxalic acid, and details of which may be found in Crookes' *Select Methods*, page 249. For another method see Fres., § 165, 4, a, also § 165, 7, a.

Note 2. — On the other hand, some of the antimony and lead may refuse to dissolve and remain with *Residue a*, in which case proceed as follows : after igniting and weighing the $SnO_2 + Pb$? $+ Sb$? fuse with Na_2CO_3 and sulphur in a porcelain crucible. Dissolve in warm water and filter from the residue of PbS, which may be treated with HNO_3 in a porcelain crucible and weighed as $PbSO_4$. To the alkaline solution add slight excess of HCl and collect precipitate of $SbS_3 + SnS_2 + S$ on filter ; dry and remove excess of S by washing with CS_2, transfer to porcelain capsule, oxidize with HNO_3, evaporate to dryness, fuse with NaHO in a silver dish, dissolve the mass in a mixture of three volumes of alcohol and one of water, and filter from the antimoniate of sodium. For details, see Fres., § 165, 4, a To the solution containing stannate of sodium, add HCl, saturate with H_2S, and treat the precipitated SnS_2 as usual. See Fres., § 126, 1, c, and § 91.

Consult article on the Estimation of Antimony, by E. H. Bartley, in *American Chemist*, Vol. V, page 436; also paper by Dr. Clemens Winkler, in Fresenius' *Zeitschrift fur Analytische Chemie*, Heft 2, 1875.

Analysis No. 15. — Zinc Ore. Determination of Zinc.

Pulverize finely; heat about 2 grms. ore with 10 c.c. HCl + 5 c.c. HNO_3 + 10 c.c. H_2SO_4 in a flask. Boil till fumes of H_2SO_4 appear. Cool; add H_2O carefully, warm and filter, wash thoroughly.

Residue a.
SiO_2 + $PbSO_4$, $CaSO_4$, etc.
Test for zinc, and, if found, treat again with acid.

Solution a.
Nearly neutralize with cryst. Na_2CO_3, dilute, add $NaC_2H_3O_2$ (about 5 grms.), boil ten minutes and filter hot. Wash hot by decantation. See Note 1 below, and Fres., § 113, 1, d.

Precipitate b.
Fe_2O_3 + Al_2O_3 as basic acetates.

Filtrate b.
Add Br water to the liquid, and digest for some time. Repeat so long as MnO_2 is precipitated. Fres., § 159, 4, a.

Precipitate c.
MnO_2 + xH_2O.

Filtrate c.
Expel Br by boiling; add a little $HC_2H_3O_2$, saturate with H_2S gas, wash the ZnS with H_2S water on the filter, carefully covering the funnel with a watch-glass. Cf. Fres. § 108, 1, b. See Note 2.

Filtrate d.
Examine carefully for Zn. and if present repeat as with Filtrate c.

Precipitate d.
Dissolve on filter with warm dilute HCl, add a very little $KClO_3$, boil to oxidize the ZnS, filter from S if necessary. and add Na_2CO_3. Fres., § 108, 1, a. and § 77. Wash, dry, ignite, and weigh as ZnO.

Note 1.—For the precipitation of Fe as basic acetate, the solution must be very carefully neutralized with crystallized Na_2CO_3, ending with a dilute solution of Na_2CO_3, and striking as deep brown-red a color as possible. Cf. Note 11, Analysis No. 11.

Note 2.—For properties of ZnS and various methods of determining Zn, see article by Hugo Tamm in American Chemist. Vol. II, p. 298.

Analysis No. 16. — CHROMIC IRON ORE.   SCHEME I.

May contain FeO, Al$_2$O$_3$, Cr$_2$O$_3$, Mn$_2$O$_3$, CaO, MgO, SiO$_2$(TiO$_2$).

Fuse 0.5 grms. ore, ground to an impalpable powder, in a large platinum crucible with 6 grms. KHSO$_4$ for twenty minutes; add H$_2$SO$_4$ from time to time, and fuse again at a higher temperature. Add 3 grms. pure Na$_2$CO$_3$ and 2 grms. NaNO$_3$, adding the latter in small portions at a time during an hour, at red heat, then heat fifteen minutes to bright redness. A little KHO added to this fusion facilitates it. Cool, remove the mass from the crucible with hot water, filter hot, and wash the residue. Fres., § 160, 8, *a*.

Residue a.	*Filtrate a.*
Digest with HCl, and filter from residue. The HCl solution is rejected. If much undecomposed ore remains, fuse again as before.	Evaporate with excess of NH$_4$NO$_3$ on a water-bath nearly to dryness, and heat until all free NH$_4$HO is expelled. Add H$_2$O, digest, and filter. Fres., § 160, 8, *a*, 77.

Residue b.	*Filtrate b.*
Al$_2$O$_3$ SiO$_2$(TiO$_2$) also Cr$_2$O$_3$, Mn$_2$O$_3$. Re-fuse and treat as before. Add second filtrate to *filtrate b.*	Boil with HCl and alcohol, expel excess of alcohol; when fully reduced add NH$_4$HO, boil, filter hot, wash thoroughly by decantation. Fres., § 166, 1, *a.*

	Filtrate c.	*Precipitate c.*
	Reject if colorless, or contains no Cr$_2$O$_3$.	Cr$_2$O$_3$. Dry, ignite, and weigh. Fres., § 76.

Analysis No. 16.— CHROMIC IRON ORE. SCHEME II.

Pulverize very finely, take 0.5 grm., fuse as in Scheme I. Dissolve in water and filter.

Residue a.

Fe_2O_3, Al_2O_3, undecomposed ore. etc. Treat with HCl, digest, filter, and wash.

Filtrate b.
Fe_2O_3. etc. Reject.

Residue b.
If not very small in quantity, must be re-fused as before and added to filtrate a.

Filtrate a.

Contains Na_3AlO_3, Na_2CrO_4, Na_2MnO_4, Na_2SiO_3, etc. Add $(NH_4)_2CO_3$, and heat nearly to boiling.

Residue c.
$Al_2O_3SiO_2$, etc.

Filtrate c.
Na_2CrO_4 solution must be relow. Neutralize with HNO_3. and add a neutral solution of $Hg(NO_3)_2$. Wash the precipitate with dilute solution of $Hg(NO_3)_2$. Dry, ignite the $HgCrO_4$, and weigh the Cr_2O_3 resulting.

For other methods, see Fresenius' *Quant. Analysis*, § 106, 2 *d.* Consult also paper by E. F. SMITH, in *American Journal of Science*, [3] xv, p. 198.

Analysis No. 17. — PYROLUSITE.

Determination of MnO_2.

Employ Fresenius and Will's method as described in Fres. *Quant. Analysis,* edition of 1881, pages 705–709, § 203, A. See also Mohr's *Titrirmethode* § 215, pp. 617–638 (ed. 1874).

Take 3.955 grms. of ore, and use *Geissler's* carbonic acid apparatus if available.

Consult also the following article: "On the Estimation of Peroxide of Manganese in Manganese Ores," by E. Scherer and G. Rumpf, *Chemical News,* American Reprint, Vol. VI, page 82, February, 1870.

Analysis No. 18. — FELDSPAR.

A. — Determination of Alkalies.

Prof. J. Lawrence Smith's method. See *Am. J. Sci.* [3] I, 269. Also Fres., § 140, II, b, γ.

Pulverize well in an agate mortar. Weigh out one grm. of the silicate. Mix well in an agate mortar, *first,* with about one grm of NH_4Cl (pure enough to sublime without residue), and, *secondly,* with about eight grms. C. P. precipitated $CaCO_3$; add the latter in three or four portions, mixing well after each addition. Transfer the mixture by means of glazed paper to a platinum crucible.

Apply the heat of a Bunsen burner to the upper portion of the crucible first and gradually carry the flame toward the lower part, until the NH_4Cl is completely decomposed, which ensues in four or five minutes. Then heat before the blast-lamp, not too intensely, for thirty to forty minutes.

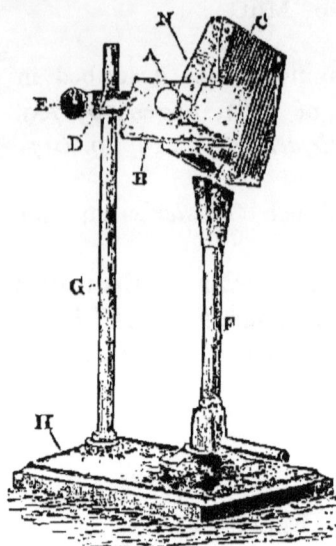

FIG. 5.

This operation is greatly facilitated by using a special apparatus devised for the purpose by Prof. J. Lawrence Smith, and represented in Fig. 5.

The stand H supports on its rod· G a cast-iron plate B perforated by a hole large enough to admit the somewhat elongated crucible A; the bottom of the crucible projects within the sheet iron chimney C which is held in its place by the hook N. When heat is applied to the bottom of the crucible by the flattened burner F the decomposition proceeds regularly and is completed in about one hour.

Cool the crucible, place it in a porcelain casserole, and digest the semi-fused mass with boiling water until thoroughly disintegrated. This may take some hours. Then filter from the residue (SiO_2, Fe_2O_3, Al_2O_3, Mn_2O_3(?), CaO, etc.), and wash well with about 200 c.c. of water. All the alkalies of the silicate are converted into chlorides and are now in the water solution. Add to this solution NH_4HO and $(NH_4)_2CO_3$ with a few drops of $(NH_4)_2C_2O_4$. Evaporate without filtering, on a water-bath, to about 50 c.c., add a little NH_4HO, and filter through a small filter (No. 2) into a weighed platinum dish. Evaporate to dry-

ness on a water-bath, ignite *very gently* to drive off a little
NH_4Cl, and weigh. If the residue is not perfectly soluble
in water, and quite white, dissolve, filter off, evaporate,
ignite, and weigh again. This gives the weight of the
$KCl + NaCl$.

Next determine the K, either by separating it with $PtCl_4$
and alcohol in the usual manner, or by gravimetric or volumetric estimation of the total Cl in the weighed chlorides.
For calculation, see Fres., page 841, 3, *a*. Consult also
Crookes' *Select Methods*, pages 13 and 14.

B. — Determination of SiO_2, Al_2O_3, Fe_2O_3, CaO, and MgO.

Fuse two grms. mineral with six grms. K_2CO_3 + six
grms. Na_2CO_3. Moisten with water, digest, add excess of
HCl, evaporate to dryness, expel HCl in air-bath, add
water and HCl, and filter from SiO_2. Continue exactly as
in *Analysis No. 7*.

Appendix to Analysis No. 18. — ANALYSIS OF SOLUBLE SILICATES.

May contain SiO_2, Al_2O_3, FeO, CaO, MgO, Na_2O, K_2O, H_2O. Pulverize, weigh out four grms., moisten with water in casserole, add conc. HCl, evaporate to dryness on water-bath. Dry in air-bath at 100°-115° C. Moisten with HCl, add water, digest, and filter.

Residue a.

SiO_2, dry, ignite before the blast lamp, and weigh.

Solution a.

Oxidize the FeO if necessary, dilute to 400 c.c., and divide into two equal portions.

Solution a¹.

200 c.c. Determine Al_2O_3, Fe_2O_3, CaO and MgO, exactly as in *Analysis No.* 7, Dolomite.

Solution a², 200 c.c.

Add solution of $Ba(HO)_2$ in excess and filter. Fres., §153, B, 4, a, a.

Residue a.

Al_2O_3, Fe_2O_3, MgO, etc. Reject. [Fe may be determined here volumetrically.]

Solution b

Add $(NH_4)_2CO_3$, boil and filter.

Residue c.

$CaCO_3$, $BaCO_3$, etc. Reject.

Filtrate c.

Add HCl cautiously. Evaporate to dryness, and heat gently over Bunsen burner until all NH_4Cl is expelled. Dissolve residue in water, filter into weighed dish. Evaporate, dry, ignite, and weigh as NaCl+KCl. Dissolve in water and determine K directly as K_2PtCl_6, or indirectly by estimation of Cl, in the mixed chlorides. See Fres., p. 841, 3, a, for calculation.

Analysis No. 19.— IRON SLAG.

To be determined: SiO_2, FeO, MnO, Al_2O_3, CaO, MgO, S, P_2O_5.

Pulverize finely; weigh out exactly five grms.; mix on glazed paper, by means of a horn spatula, with fifteen grms. anhydrous Na_2CO_3 and fifteen grms. K_2CO_3, together with one grm. $NaNO_3$. These fluxes need not be accurately weighed. Put one-third the mixed slag and fluxes into a two-ounce platinum crucible, and heat over a Bunsen burner until by settling down room is made for more. Heat twenty minutes or more before the blast-lamp. Cool suddenly, place in a casserole, and treat with boiling water until thoroughly disintegrated. Remove the crucible and add excess of HCl little by little, avoiding loss of liquid by violent effervescence; evaporate to dryness on water-bath, expel HCl completely by drying (not above 115° C.) in an air-bath.

Moisten with water, add HCl, digest, and proceed as per scheme on following page.

Filter the solution obtained as directed on page 60.

Residue a.	Filtrate a.		
	Dilute to 500 c.c. and divide into three portions.		

Filtrate a. Dilute to 500 c.c. and divide into three portions.

Residue a.	Solution a³. 300 c.c.	Solution a². 100 c.c.	Solution a¹. 100 c.c.

Residue a. SiO₂. Dry, ignite, and weigh.

N.B. — Check SiO₂ by fusing 1 grm. slag as above and following details there given.

Solution a³. 300 c.c. Cool, nearly neutralize (in a large flask) with cryst. Na₂CO₃ + 10H₂O, add about 15 grms. NaC₂H₃O₂, dilute to about 2 litres, heat to boiling, and filter hot. Wash well. See *Note* 11 to *Analysis No. 21.*

Precipitate b.	Filtrate b.
Fe₂O₃ + Al₂O₃ as basic acetates; also P₂O₅. Wash, dissolve in strong HCl, and divide into two unequal portions.	Mn, Ca, Mg. Proceed exactly as in "*Filtrate g*" in *Analysis No. 21,* Scheme II. Omit, however, the treatment with Br. if Mn is known to be wanting.

Solution b¹.	Solution b².
To determine P₂O₅, proceed exactly as with "*Solution g¹*" in *Analysis No. 21.*	To determine Al₂O₃, proceed exactly as with "*Solution g²*" in *Analysis No. 21.*

Solution a². 100 c.c. Add excess of NH₄HO, wash, dissolve in H₂SO₄, and determine the Fe with K₂Mn₂O₈. Cf. *Note* 18. *Analysis No. 21.*

Solution a¹. 100 c.c. Add BaCl₂ and treat the BaSO₄ in the usual manner. Report S.

Filtrate from BaSO₄ may be kept for determining Fe in case of accidents.

Note. — If the slag contains much manganese, the solution of the fused mass will be strongly colored green from the formation of sodium and potassium manganates; on boiling this solution it becomes of a violet color in accordance with the following reaction: $3K_2MnO_4 + 3H_2O = MnO_2H_2O + K_2Mn_2O_8 + 4KHO.$

On adding HCl to the permanganate solution it loses its color owing to following reactions:

$K_2Mn_2O_8 + 8HCl = 2(MnO_2H_2O) + 2KCl + 2H_2O + 6Cl,$ and $MnO_2H_2O + 4HCl = MnCl_2 + 3H_2O + 2Cl^2$

Analysis No. 20.—HEMATITE.

Determination of Fe, SiO_2, S, and P.—Pulverize very finely, and weigh out on a watch-glass exactly 5 grms., mix on glazed paper with 25 grms. pure Na_2CO_3+2 grms. $NaNO_3$, and fuse in 2 oz. platinum crucible. (Consult *Note* 2. *Analysis No.* 21.). Cool suddenly, place in a casserole, and treat with boiling water until thoroughly disintegrated. Remove the crucible (*Note* 3, *Analysis* 21), and add carefully excess of HCl. Evaporate to dryness on water-bath, expel HCl in air-bath at 110°–115° C. Add HCl, digest, dilute, and filter.

Residue a.	*Filtrate a.*		
SiO_2. Dry, ignite thoroughly, and weigh. If not white after ignition repeat the fusion and treat as before.	Dilute to 500 c.c., and divide into 3 portions.		
	Solution a¹. 100 c.c. Determine S as $BaSO_4$, in the usual manner. The filtrate from $BaSO_4$ may be reserved for duplicating the Fe.	*Solution a².* 100 c.c. Add excess of NH_4HO, wash to remove NH_4Cl, do *not* bring on filter, dissolve in H_2SO_4, reduce and determine Fe by $K_2Mn_2O_8$. See *Note* 18, *Analysis No.* 21.	*Solution a³.* 300 c.c. Add NH_4HO in excess, and proceed for determination of P exactly as with "*solution g*" in *Analysis No.* 21.

Quick Method for the Determination of Iron only.—Sample, pulverize, fuse 1 grm. $Na_2CO_3 \times 1$ grm. $NaNO_3$, about 20 minutes. Plunge crucible while hot into cold water in a casserole; boil, and after removing crucible neutralize carefully (CO_2 escapes) with conc. H_2SO_4; add excess of acid, filtrate, if much remains undissolved, dilute to 500 c.c., divide in halves, reduce with amalgamated zinc and platinum foil, and *titrate* with $K_2Mn_2O_8$, as usual. See for details of the latter steps, *Notes to Analysis No.* 21, *Scheme II.*

Analysis No. 21.—TITANIFEROUS IRON ORE. SCHEME II.*

Prepare an average sample for analysis. (*Note* 1.) Pulverize in an agate mortar to an impalpable powder. Make a qualitative examination for H_2O—TiO_2—Cu—As and Cr. If Cu—As or Cr are present, see Scheme I.

To be determined : TiO_2—SiO_3—Fe—Al_2O_3—Mn—CaO—MgO—S—P—H_2O.

Weigh out exactly 5 grammes, mix with 20 to 30 grammes Na_2CO_3, and 2 to 5 grammes $NaNO_3$, and fuse in a platinum crucible. (*Note* 2.) Cool suddenly, place in a casserole and treat with boiling water until the mass is thoroughly disintegrated. (*Note* 3.) Filter and wash with hot water. (*Note* 4 and *Fres.*, § 160, S, a.)

1. Water Solution.

It must be perfectly clear but may be colored, and may contain SiO_3—SO_3—P_2O_4—and Al_2O_3. Add carefully an excess of HCl. evaporate on water-bath to dryness, heat in an air-bath at $100°$ C. to $115°$ C. till odor of HCl is no longer perceptible. (*Note* 5.) Moisten residue with HCl, add water, digest, filter, and wash hot.

Residue a.	Filtrate a.		
	Dilute to 500 c.c. and divide into three portions.		
	Solution a^1—300 c.c.	Solution a^2—100 c.c.	Solution a^3—100 c.c.
SiO_2. To be added to and re-fused with *Residue b.*	Put into a large flask to be afterwards combined with *Filtrate f.*	Add BaCl, and determine H_2SO_3 as $BaSO_4$ (*Note* S and *Fres.*, § 132.)	Add to *Solution b²*, as a little Fe often enters the water solution.

*Scheme I may be found in *American Chemist* Vol. I, p. 123. Both schemes are modifications of one originally drawn up by Dr. C. F. Chandler (See preface to this work.)

2. Insoluble Residue.

It may contain SiO_2—TiO_2—P_2O_5—Fe_2O_3—Al_2O_3—Mn_2O_3—CaO—MgO (and Pt from the crucible). Dry the residue on the filter, transfer to a casserole, burn the filter and add the ashes. Moisten with water. add conc. HCl, evaporate to dryness, heat till HCl is expelled; add conc. HCl, then water. (Note 9.) Digest with occasional stirring, filter, and wash. (Fres., § 140.)

Residue b.

Contains SiO_2—TiO_2. etc. Combine with Residue a, fuse with 5 parts Na_2CO_3, remove fused mass from crucible with hot water, acidulate with HCl, evaporate to dryness not above 100° C. (Fres., § 140. II, a.) Add HCl and boiling water, filter and wash.

Residue c.	Filtrate c.
Dry on funnel. fuse with 6 parts $KHSO_4$; dissolve in about 300 c.c. cold water. filter and wash cold.	Add this filtrate and washings to Filtrate b.

Residue d.	Filtrate d.
Dry, ignite and weigh as SiO_2.	Dilute to 500 c.c. and divide into 3 portions. d^1, d^2 and d^3.

Filtrate b or Hydrochloric Acid Solution.

Combine with Filtrate c. Dilute to 500 c.c. and divide into 3 portions.

Solution b^1—300 c.c.	Solution b^2. 100 c.c.	Solution b^3. 100 c.c.
Saturate thoroughly with H_2S gas and filter from the PtS_2 and S.	Combine with Solutions a^3 and d^2, add excess of NH_4HO. wash by decantation twice. dissolve in H_2SO_4, dilute largely; part-ly neutralize with Na_2CO_3, saturate with H_2S gas, boil 5 to 7 hours, adding water and H_2S water and H_2S water and wash. (For ppt. h and filtrate k see next page.)	To be combined with solution d^3 and re-served for accidents.

Filtrate e.

Boil with $KClO_3$ to oxidize FeO.

Filtrate f.

Cool, combine with Solutions a^1 and d^1 in large flask. Add Na_2CO_3 almost to neutralization and about 20 grammes sodic acetate. dilute to about 2.5 litres, and boil ten minutes. (Note 11 and Fres., § 113, 1, d.)

Precipitate g.	Filtrate g.
Contains Fe_2O_3, and Al_2O_3 as basic acetates and perhaps P_2O_5 and TiO_2. Wash thoroughly, dissolve in strong HCl and divide into 2 portions, g^1 and g^2.	Contains Mn—Ca and Mg. Concentrate to small bulk, add Br, digest until excess of Br is expelled, filter and wash. (Fres., § 109, 1, d.) (For ppt. h and filtrate h see next page.)

Solution d¹	Solution d²	Solution d³	Solution g¹ 3.	Solution g² 3.	Precipitate h	Filtrate h. Contains Ca and Mg.	Precipitate k	Filtrate k.
300 c.c. Add to Filtrate f.	100 c.c. Add to Solution b².	1.0 c.c. Add to Solution b².	Add NH_4HO in large excess, wash twice by decantation and re-dissolve in conc.HNO_3. Boil down to small bulk, add NH_4NO_3 and 50 c.c. $(NH_4)_2MoO_4$ +HNO_3). Warm and set aside 24 hours. Filter, test filtrate, wash with the diluted precipitant (½ and §113.) Dissolve ppt. in original beaker, filter through same filter, add "magnesia mixture" and determine P_2O_5 as usual. (*Fres.*, § 134 I., 6, β.)(Compare *Note 12*.)	Add $(NH_4)HO$ in excess to precipitate $Fe_2O_3+Al_2O_3+P_2O_5$ ($+TiO_2$). Boil till all free NH_3 is expelled, filter, wash thoroughly, dry, ignite and weigh. (*Fres.* § 105 and §113, 1, *a*, and *Note a*, and §¹.) From this weight deduct P_2O_5 found in g¹, together with the Fe_2O_3 calculated from k² and TiO_2 found in k; difference $=Al_2O_3$.	Dissolve in HCl on filter and wash. Boil, add Na_2CO_3 in excess, boil, filter and wash. Dry ignite and weigh as Mn_3O_4. (*Fres.*, § 109, 1, *a*.) (Compare *Note 16*.)	Add NH_4HO NH_4Cl and $(NH_4)_2C_2O_4$, let stand 12 hours, filter and wash hot. Precipitate i. Dissolve ppt. in HCl. re-precipitate with NH_4HO filter and wash (add filtrate to filtrate i.) Dry ppt., burn filter separately. moisten with H_2SO_4 in crucible and ignite. Weigh as $CaSO_4$. (*Fres.*, §101,2,6,c.) — Filtrate i. Add NH_4HO and Na_2HPO_4, let stand 12 hours. filter. wash,dry and ignite. Weighas $Mg_2P_2O_7$. (*Fres.*, § 104, 2.)	Consists of TiO_2. Dry, ignite and weigh. If dark colored. fuse with 6 pts. $KHSO_4$, dissolve in cold water, filter and reprecipitate by boiling, filter, wash, ignite and weigh again as TiO_2. (*Fres.*, § 107.) See *Sundry Suggestions*, No. 4.	Expel the H_2S by boiling with $KClO_3$, concentrate filter, dilute to 500 c.c. and divide into 2 portions. Solution k². Reduce the Fe_2O_3 by amalgamated Zn and Pt foil and determine Fe volumetrically by Marguerite's process. (*Fres.*, § 112.) i., a.) Solution k³. Treat in exactly the same way as *solution k¹* and average the results. (*Compare Note 18.*)

Special Determination. Determine H_2O in 1 grm. of ore by direct weight. (*Fres.* § 16.)

Notes to the Preceding Scheme.

Note I. SAMPLING THE ORE.—Break up in an iron mortar forty or fifty pounds into pieces that will pass through a tin sieve with half-inch holes. Thoroughly mix the fine and coarse. Break up about ten pounds of average quality, so that it will pass through a tin sieve with quarter-inch holes. Mix well, take one pound, and pulverize in the iron mortar until it will pass through a brass sieve of 60 meshes to the linear inch. Mix well, take out about 50 grammes, pulverize in agate mortar, pass through muslin bolting cloth, and put into a small bottle, tightly corked, for analysis and special determinations. It is yet necessary that every portion of this required for the main analysis or a special determination should be further pulverized, as needed, in an agate mortar, to an *impalpable powder.*

Note 2. PRELIMINARY FUSION.—Thoroughly mix the ore and its fluxes on glazed paper, put about a third of the mixture in a two-ounce platinum crucible, the lower portion of whose interior surface has been previously lined with a thin layer of Na_2CO_3, and heat over a common Bunsen burner with strong flame until the greatest violence of the effervescence has ceased. Then add and treat the two-thirds remaining successively and with the same precaution. Finally, heat strongly over the blast-lamp until the mass is in complete and quiet fusion, adding a little more Na_2CO_3, should it not readily fuse. The time required for this fusion varies from 30 to 50 minutes.

Certain highly aluminiferous ores obstinately resist this method of attack; in such cases mix with the flux a known weight (two or three grammes) of chemically pure precipitated silica which has been strongly ignited just before weighing. The amount of silica added is afterwards deducted from the total amount found in *Residue d.*

Note 3. REMOVAL OF THE FUSED MASS.—Let the crucible cool until just below red heat, then chill it suddenly by plunging it into cold water contained in a porcelain casserole, lay the crucible on its side and digest with boiling water. The fused mass will generally become detached from the crucible and come out in a cake. Then remove the crucible, wash it, treat in a small beaker with a little conc. HCl to remove any adhering particles of the mass, and add this solution to that of the INSOLUBLE RESIDUE (2). Should any portion of the fused mass, thicker than a film, obstinately resist solution in the hot water, it ought to be removed only by patience and long boiling; and no attempt should be made either to *dig* it out or to dissolve it in HCl; lest by the formation of *Aqua Regia* or free Cl (in the presence of $NaNO_3$, or Mn_2O_3) the crucible be attacked and injured.

Note 5. SEPARATION OF SiO_2.—In order to render the SiO_2 entirely insoluble, it must be perfectly dehydrated. The evaporation should be carried to dryness, the residue heated until odors of HCl can no longer be detected, and the mass is hard and crumbly. Since the residue is to be re-fused with *Residue b*, the drying may be completed, at a temperature somewhat higher than 100° C., in an air-bath.

Note 8. PRECIPITATION OF $BaSO_4$.—Avoid the addition of a large excess of $BaCl_2$ solution. Add only 5 c.c. at first, and then after complete subsidence of precipitate, add a few drops to determine if any H_2SO_4 remains unprecipitated, etc. Then proceed as in Fres., § 132, I, 1. After decanting the clear supernatant liquid, boil the precipitate with water, allow to subside, decant, filter, and wash with hot water. These precautions are necessary to dissolve out any other salts of barium, which are always carried down on the first precipitation. If the precipitate of $BaSO_4$ is dark colored after ignition, dissolve in the crucible in hot conc. H_2SO_4, pour into cold water, and collect the precipitate as before.

Note 9. SEPARATION OF SiO_2.—Evaporate as in *Note* 5. Then add HCl quite freely and warm for some time before adding any water, as the high heat may have produced anhydrous Fe_2O_3, forming an oxychloride which is very slow to dissolve, especially in dilute acid. Should the acid already added be too dilute, concentrate by evaporation, add conc. HCl, and digest at a moderate heat.

Note 11. PRECIPITATION OF THE BASIC ACETATES.—*Filtrate f* combined with *Solutions a'* and *d'* must be very carefully neutralized with sodium carbonate. (If ammonium carbonate were used, bromide of nitrogen might form in *Filtrate g*.) To neutralize the greater portion of the acid use crystallized sodium carbonate, and complete the neutralization with a very dilute solution of the carbonate, adding it drop by drop, agitating to dissolve the precipitate, until the liquid assumes a deep mahogany-red color. If a permanent precipitate forms, add a little hydrochloric acid, and repeat as above. Then dilute the solution to about 1 litre for each gramme of the sesquioxide present, add about 20 grammes sodium acetate dissolved in a small quantity of water, and heat the whole to boiling.

It is sufficient to boil from ten to fifteen minutes for the complete precipitation of the acetates. The filtering should be done rapidly on a ribbed filter, keeping the fluid hot, and disturbing the settled precipitate as little as possible. When available the Bunsen pump may here be used with advantage. After the supernatant fluid has been poured through the filter, throw on the precipitate and wash it with boiling water containing a little sodium acetate. Should any *basic acetate* separate upon concentrating the filtrate, add some sodium acetate, boil, filter, dissolve the precipitate in HCl, and unite to the solution of the main body.

In boiling *Filtrate e* with $KClO_3$ to oxidize FeO, be careful to decompose the whole of the chlorate by heating with excess of HCl.

Note 12. DETERMINATION OF P_2O_5.—To remove the HCl in *Solution g* add NH_4HO in large excess, wash the precipitates of ferric hydrate and ferric phosphate by decantation two or three times, and redissolve in hot conc. HNO_3. Evaporate this solution down to small bulk (150 c.c. to 100 c.c.), partially neutralize with NH_4HO, and add about 50 c.c. of solution of ammonium molybdate in nitric acid. If the solution is very acid, ammonium nitrate is formed by the partial neutralization as above, otherwise add a small quantity of the salt. Warm the solution, do not boil, and let stand 24 hours or more. Then filter from the yellow granular precipitate of ammonium phospho-molybdate without bringing it all on the filter, and wash the precipitate with a solution prepared by mixing 100 parts of the precipitant with 20 parts of HNO_3 (sp. gr.$=1.2$) and 80 parts of water. Dissolve the yellow precipitate by pouring a small quantity of dilute NH_4HO through the filter into the original beaker, and determine the phosphoric acid in the ammoniacal solution by means of magnesia mixture (5 c.c.) in the usual manner. Magnesia mixture is preferably made with magnesium chloride. If the crystalline ammonio-magnesium phosphate falls mixed with flocculent magnesium hydrate, add HCl until dissolved and reprecipitate with NH_4HO.

Reserve the filtrate and washings of the yellow precipitate, and test for phosphoric acid by adding a little more of the ammonium molybdate solution, heating and allowing to stand 12 hours. If a yellow precipitate forms, pour through a separate filter, dissolve in dilute NH_4HO and add to the ammoniacal solution.

If the yellow precipitate first obtained was not suf

ficiently washed, a red residue of oxide of iron may remain on the filter, in which case pour dilute HNO_3 upon it, allow it to pass into the ammoniacal solution, acidulate that with HNO_3, warm, add more of the precipitant, and set aside as before; filter and wash several times with the diluted precipitant, then dissolve the precipitate on the filter and that adhering to the beaker in as little dilute NH_4HO as possible.

The yellow granular precipitate of ammonium phospho-molybdate is not sufficiently constant in composition to admit of directly weighing it in exact analysis; it is therefore dissolved in NH_4HO and the phosphoric acid thrown down with magnesia mixture as just detailed. According to Nuntzinger's analysis, after drying at $100°$ C., it contains

$$3.577 \text{ per cent. } NH_4HO$$
$$3.962 \quad `` \quad P_2O_5$$
$$92.461 \quad `` \quad MoO_3$$

$$\overline{100.000}$$

Lipowitz says the precipitate dried at $20°$ to $30°$ C. contains 3.607 per cent. of P_2O_5, and Eggertz 3.7 to 3.8 per cent. P_2O_5. When dried at $120°$ C., Sonnenschein found about 3 per cent. For properties of this precipitate see also Fres., § 93, i, *foot-note.* Consult also Finkener's paper in *Bericht d. d. chem. Ges* XI, p. 1638 (1878), and *Chem. News*, XLVII, p. 66 (1883).

Note 13. WASHING OF $Fe_2O_3 3H_2O$.—Wash this precipitate by boiling up with water and decanting until the wash water shows very little alkaline reaction with litmus paper, and gives very little precipitate with solution of $AgNO_3$. Then transfer to filter, and wash thoroughly with boiling water.

Note 16. DETERMINATION OF MN.—(Gibbs' process, *Am*

Jour. Sci. [2] XLIV, p. 216.) To the HCl solution add NH_4HO in excess and solution of Na_2HPO_4 in large excess. Then add dilute H_2SO_4 or HCl until the white precipitate redissolves, heat to boiling, and add NH_4HO in excess. Digest near the boiling point about an hour, when the precipitate, at first white and gelatinous, becomes rose-colored and forms crystalline scales. Filter and wash with hot water. If tinged red, redissolve the precipitate in dilute HCl, and repeat the process. On ignition the precipitate is converted into $Mn_2P_2O_7$, a nearly white powder.

If Zn is present, it must first be separated as in Scheme I, *Am. Chem.*, Vol. I, p. 323.

Note 18. Volumetric Determination of Fe.—Put *Solution k'*, which must be completely free from the $KClO_3$, used to oxidize *Filtrate k*, into a wide-mouthed reduction bottle holding about 250 c. c. Carefully let down into the bottle a lump of amalgamated zinc, free from iron, and a strip of platinum foil resting upon it, add about 10 c. c. conc. H_2SO_4, cover with a watch-glass and set aside over night. To ascertain if the reduction is complete test the solution with ammonium sulpho-cyanide, which should give only a trace of pink color.

Then introduce into a flask holding about 200 c. c., and fitted with a Krönig valve, exactly 0.2 gramme iron pianoforte wire, add dilute H_2SO_4, and heat until complete solution of iron. Cool the flask, pour and wash out the contents of the flask into a large beaker containing about 400 c. c. cold water, add a little concentrated H_2SO_4 and titrate with a solution of $K_2Mn_2O_8$ (13 grms. in 2 litres water) to determine its strength. Repeat, and average results.

Now pour and wash out the contents of the reduction-bottle into a large beaker, add conc. H_2SO_4, and titrate with the standard $K_2Mn_2O_8$ as before. If the HCl was not properly removed from *Solution b[2]* the dark brown-red ferric chloride formed will interfere with the end reaction of the permanganate. In such a case reprecipitate with NH_4HO, wash thoroughly, and proceed as with *Solution k'*.

Treat *Solution k'* in exactly the same manner, and average the results. Cf. *Analysis No.* 3, C. III.

For method of repeating the titration in the same solution, see Crookes' *Select Methods*, p. 74.

SUNDRY SUGGESTIONS.—I. *Solution a[3]* may be used for duplicating the determination of S, provided the absence of Fe is proved by the proper tests. Duplicate determinations of Ca and Mg can be made, if desired, in the filtrate from the precipitate formed by ammonium hydrate in *Solution b[2]*, provided this precipitate be thoroughly washed.

2. Duplicate determinations of Ti and of Fe can be made in *Solution b[3]*; the Fe can also be estimated volumetrically by dissolving in acid the weighed precipitate resulting from the treatment of *Solution g[2]*. In the latter case, however, the presence of TiO_2 will impair the results.

3. The purity of the SiO_2 obtained in *Residue d* may be tested, after weighing, by heating with fluoride of ammonium and concentrated sulphuric acid in a platinum crucible, whereby all the SiO_2 is expelled and is determined by the loss in weight, the residue being TiO_2 probably colored by Fe.

4. In fusing *Residue c* or *Precipitate k*, hydro-sodium sulphate may be substituted for $KHSO_4$, but since the former contains water of crystallization it should be heated until the water is expelled before using in fusions. In either case avoid expelling the whole of the H_2SO_4, or if the mass is heated to redness, partially cool, add conc. H_2SO_4 and heat again at a lower temperature. In this

way the TiO$_2$ will be held in solution by the excess of acid, and the resulting acid sulphate will dissolve out readily.

For *Special Determinations* see NOTES TO SCHEME I in *American Chemist*, Vol. I, pp. 323 *et seq.*

REACTIONS.—A full discussion of the many and complex reactions which take place in the preceding scheme for the analysis of iron ores is superfluous.

We add a few remarks and equations which may serve to throw light upon some points.

A.—The action of potassium permanganate on ferrous sulphate has already been formulated in connection with the notes to *Analysis No. 3*. This action, however, may be regarded as taking place in two stages, as follows:

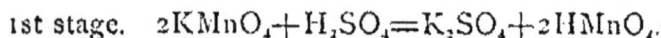

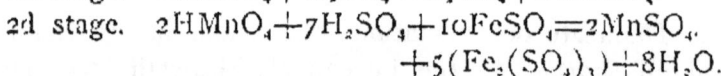

1st stage. $2KMnO_4 + H_2SO_4 = K_2SO_4 + 2HMnO_4$.
2d stage. $2HMnO_4 + 7H_2SO_4 + 10FeSO_4 = 2MnSO_4$.
$$+ 5(Fe_2(SO_4)_3) + 8H_2O.$$

Solution b² is treated with excess of NH$_4$HO and the precipitate dissolved in H$_2$SO$_4$ in order to remove the larger part of the HCl which might vitiate the results of the titration as indicated in Note 18. The presence of HCl is injurious also because it exerts a reducing action on the permanganate as shown in the equations following:

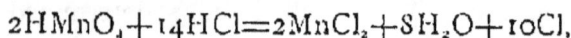

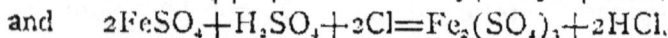

$$2HMnO_4 + 14HCl = 2MnCl_2 + 8H_2O + 10Cl.$$
and $\quad 2FeSO_4 + H_2SO_4 + 2Cl = Fe_2(SO_4)_3 + 2HCl.$

B.—When KClO$_3$ is employed in acid solution as an oxidizing agent (as in the case of *Filtrate e*), the reaction which takes place depends upon the acid used and partly upon the strength of said acid. Concentrated sulphuric acid is said to act thus:

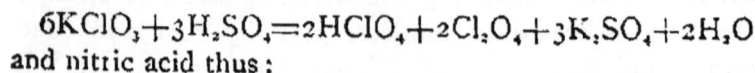

$$6KClO_3 + 3H_2SO_4 = 2HClO_4 + 2Cl_2O_4 + 3K_2SO_4 + 2H_2O$$
and nitric acid thus:

$$8KClO_3 + 6HNO_3 = 2KClO_4 + 6KNO_3 + 6Cl + 13O + 3H_2O.$$

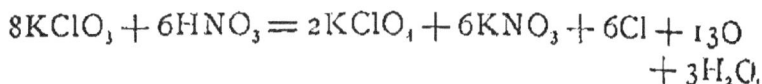

The action of hydrochloric acid on potassium chlorate is variously formulated; Böttger gives the equation (1) and Odling (2):

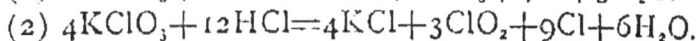

(1) $2KClO_3 + 6HCl = 2KCl + Cl_2O_3 + 4Cl + 3H_2O.$

(2) $4KClO_3 + 12HCl = 4KCl + 3ClO_2 + 9Cl + 6H_2O.$

In any of these cases the powerful oxidizing agency of $KClO_3$ is evident.

Appendix to Analysis No. 21.

A.—Method for the Estimation of Fe and Ti only.

Sample, pulverize, fuse 1 grm. ore with 3 grms. $NaFl + 12$ grms. $KHSO_4$. Dissolve in large quantity of *cold* water; if there is any considerable residue re-fuse. Neutralize with Na_2CO_3 until a slight precipitate forms, then add H_2SO_4 until the ppt. redissolves and the liquid is slightly acid. Saturate with H_2S gas, boil some hours, occasionally adding H_2S water. Filter from the precipitate of $TiO_2 + S$, dry, ignite, and weigh, if dark colored refuse, etc. To filtrate add a little $KClO_3$, boil to oxidize H_2S. Reduce the iron with amalgamated zinc and platinum foil, and titrate with $K_2Mn_2O_8$ as usual. As a result of the fusion we have

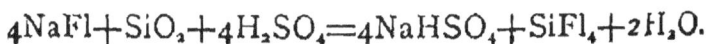

$$4NaFl + SiO_2 + 4H_2SO_4 = 4NaHSO_4 + SiFl_4 + 2H_2O.$$

B.—Flight's Method for the Separation of Iron, Alumina and Phosphoric Acid.

(*Journal of Chemical Society* (2). XIII., 592, 1875. The solution of the three substances named must contain but little free hydrochloric acid. Boil the solution two or three hours with an excess of sodium hyposulphite, and filter. Wash thoroughly.

Precipitate a.

Contains all the Al_2O_3, and most of the P_2O_5.* Dissolve in HCl, add enough NaHO to completely redissolve the precipitate formed and throw down P_2O_5 with excess of BaCl. Do not heat, but let stand a few hours covered. Wash with dilute NaHO.

Filtrate c.

Al_2O_3 in alkaline solution. Acidify with HCl and determine Al_2O_3 in the usual manner.

Precipitate c.

Dissolve the $Ba_3P_2O_8$ in HCl, add slight excess of H_2SO_4, boil and filter.

Precipitate d.

$BaSO_4$. Reject.

Filtrate d.

Combine with *Filtrate b* and determine P_2O_5 with magnesia mixture in the usual manner.

Filtrate a.

Contains all the iron and some of the P_2O_5. (If but a small amount of P_2O_5 is present in the solution, this filtrate will contain no P_2O_5 and may be rejected after careful testing.) Add NH_4HS saturated with H_2S and warm. Filter quickly, wash with H_2S water containing a few drops of NH_4Cl.

Precipitate b.

FeS. Dissolve in HCl, oxidize with HNO_3, precipitate with NH_4HO and determine as usual.

Filtrate b.

P_2O_5. Reserve to add to *Filtrate d.*

* Flight states that the P_2O_5 is carried down with the Al_2O_3 completely when the solution contains less than 45 per cent. P_2O_5.

ANALYSIS No 22.—PIG IRON.

To be determined: Iron, Combined Carbon, Graphite, Silicon, Sulphur, Phosphorus, and Manganese.

A.—Determination of Graphite, Silicon, Sulphur, Phosphorus and Manganese.

(By F. A. CAIRNS.)—Place 10 grms. of fine borings in a flask of about two litres capacity, add 25 to 35 grms. $KClO_3$, little by little, a few grms. at a time, pour in carefully and gradually concentrated HCl, using eventually about 300 c. c. Digest until the iron is completely dissolved, then pour contents of flask into a porcelain dish and evaporate to dryness on a water-bath. Moisten with HCl, add water, filter through a weighed filter, previously dried at 100° C.

Residue a.	Filtrate a.		
	Dilute to 1000 c. c. and divide into three portions as follows:		
Graphite and *Silicon.* Wash thoroughly and weigh on the filter after drying at 100° C. Then transfer to a platinum crucible and burn off the graphite; weigh the residue as SiO_2. *See Note 2.* If the residue contains iron, expel the SiO_2 by heating with NH_4Fl and H_2SO_4 and weigh again. Compare of *Analysis No. 21,* Scheme **AA**, or the Second Method. II.	*Solution a¹.* 500 c. c. For determination of *phosphorus* proceed exactly as with *Solution g¹* of *Analysis No. 21,* Scheme II.	*Solution a².* 300 c. c. For determination of *sulphur* partially neutralize with solution of Na_2CO_3, and proceed as with *Solution a²* of *Analysis No. 21,* Scheme II.	*Solution a³.* 200 c. c. For determination of *manganese* proceed exactly as with *Filtrate f* and *Filtrate g* of *Analysis No. 21,* Scheme II. *See Note 1.*

Note 1.—Care must be taken in dissolving the pig-iron in HCl-|-KClO₃ not to add the oxidizing agent all at once, nor too rapidly, otherwise some of the iron may remain unoxidized. Should a small portion of ferrous chloride remain in the solution, the subsequent precipitation of the iron as basic acetate (as in *Filtrate f, Analysis No.* 21) will be imperfect; instead of an orange red flocculent precipitate resembling ferric hydrate, the iron will fall as a brick-red pulverulent precipitate, (anhydrous ferric oxide?) which has the property of running through filters.

Note 2.—SiO_2 obtained in this manner, and dried at 100° C., contains 6 per cent. H_2O, which is expelled on ignition, and must be deducted from graphite after the SiO_2 has been determined. According to Allen (see *Chemical News*, Vol. XXIX., p. 91, Feb., 1874) the Si of the pig-iron is converted by the action of dilute HCl into *leucone*, $3SiO.2H_2O$. By fusing the mixture of leucone and graphite with KHO, the former goes into solution, and both may be estimated directly.

AA.—Determination of Graphite and Silicon.

Second Method. (EGGERTZ, *Chem. News*, XVIII., p. 232.—Mix 10 c.c. H_2SO_4 with 50 c.c. H_2O, cool, add 5 grms. fine borings, boil half an hour, evaporate one-third and cool. The reaction is as follows :

$$2Fe_4C + 8H_2SO_4 = 8FeSO_4 + C_2H_4 + H_{11}.$$

This equation, however, but imperfectly formulates the reaction, the S forming H_2S and the P forming PH_3. A large number of compounds of C and H are evolved in addition to the C_2H_4 of the equation; according to Dr.

Hahn (*Annalen der Chemie und Parmacie*, Vol. 129, p. 57, 1864) they include the following:

Gaseous.	Ethylene,	C_2H_4.
	Propylene,	C_3H_6.
	Butylene,	C_4H_8.
Liquid.	Amylene,	C_5H_{10}.
	Caproylene.	C_6H_{12}.

Liquid.	Œnanthene,	C_7H_{14}.
	Caprylene,	C_8H_{16}.
	Elaene,	C_9H_{18}.
	Paramylene,	$C_{10}H_{20}$.
	Cetylene,	$C_{16}H_{32}$.
	etc.	etc.

Next add 10 c. c. HNO_3 and boil 15 minutes.

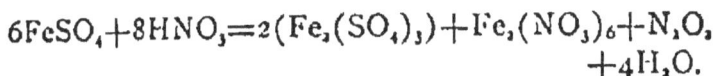

$$6FeSO_4 + 8HNO_3 = 2(Fe_2(SO_4)_3) + Fe_2(NO_3)_6 + N_2O_2 + 4H_2O.$$

Evaporate on a water-bath until vapor ceases to come off and the mass is nearly dry.

Add 75 c. c. $H_2O + 13$ c. c. HCl and boil 15 minutes; add more HCl if any Fe_2O_3 remains undissolved. Filter through a filter washed with HCl, dried and weighed; wash first with cold water until no more iron appears in washings, then with boiling water containing 5 per cent. HNO_3. Dry at 100° C., and weigh the residue consisting of $SiO_2 +$ graphite. Ignite and weigh again; the loss in weight gives the amount of graphite. Lest the residue contain something besides SiO_2 it is well to determine the latter by heating with NH_4Fl and H_2SO_4, which expels the SiO_2, in accordance with the following equation:

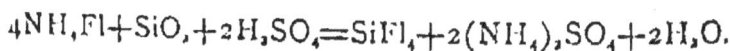

$$4NH_4Fl + SiO_2 + 2H_2SO_4 = SiFl_4 + 2(NH_4)_2SO_4 + 2H_2O.$$

The loss in weight gives the amount of SiO_2; consult, however, *Note* 2 of Δ.

ΔΔΔ.—Graphite determination according to F. A. CAIRNS.

Dissolve 5 grms. borings in dilute HCl, boil, filter, wash with hot water, then with KHO solution, then with boiling water, then with (*a*) alcohol, (*b*) ether and (*c*) hot water. Dry and transfer to flask and determine as in B.

In this process the combined carbon goes off in volatile hydrocarbons, and graphite $+SiO_2$ together with certain liquid hydrocarbons, remain. The SiO_2 is removed by the KHO, the hydrocarbons dissolve out in the alcohol and the ether, while the latter is removed at last by boiling water.

B.—Determination of Total Carbon.

A. II. ELLIOTT's modification of ROGER's Process. See *Journal of Chemical Society*, London, May, 1869; also CAIRNS' article in *Am. Chem.*, Vol. II, p. 140.

To 2.5 grms. of borings add 50 c.c. of a neutral solution of $CuSO_4$, containing one part of sulphate to 5 parts of water; heat gently for 10 minutes; the iron dissolves; copper is precipitated, and the silica, graphite, and combined carbon remain :

$$Fe+CuSO_4=Cu+FeSO_4.$$

The cupric sulphate should be as neutral as possible, in order to avoid loss of combined carbon, in the form of volatile hydrocarbons, as shown in AA.

Add 20 c.c. $CuCl_2$ (1 part of chloride to 2 parts of water), with 50 c.c. strong HCl, and heat for some time nearly to boiling, until the copper dissolves :

$$CuCl_2+Cu=Cu_2Cl_2.$$

Prepare an asbestus filter as follows : select a glass tube of about 3 to 4 cm. diameter, and 18 to 20 cm. in length. Draw out this tube to taper at one end, and place broken glass and asbestus, lightly packed, in the narrowed portion of the tube. (See *Fres.*, § 218, I, 1, p. 759.) Filter the cuprous solution through the asbestus, wash thoroughly with boiling water, and transfer contents of filter to a flask holding about 200 c.c. In making this transfer, the carbon, asbestus, and broken glass may be blown into the flask together, in order to use as little water as possible. Add to the contents of the flask about 3 grms. of CrO_3, (or if this is not available, about 5 grms. $K_2Cr_2O_7$,), and arrange apparatus as in the determination of CO_2 by direct weight, *Analysis No. 7*, note 8, II (page 34). Avoid adding more water than absolutely necessary to transfer the carbon. Add 30 c.c. to 40 c.c. concentrated H_2SO_4, little by little, shaking constantly, and closing cock of funnel-tube each time. Finally, heat gently to boiling, not allowing more than three bubbles of CO_2 gas to pass per second:

$$3C + 4CrO_3 + 6H_2SO_4 = 3CO_2 + 2Cr_2(SO_4)_3 + 6H_2O.$$

Boil one minute, attach guard tube of soda lime, and aspirate slowly, three bubbles per second. Weigh the soda-lime tube for amount of CO_2 absorbed, and calculate the amount of carbon.

Note—The carbon separated from cast-iron by treatment with sulphate of copper contains H and O, and cannot therefore be determined by weighing directly. Schutzenberger and Bourgeois assign to it the composition expressed by the formula $C_{11}3H_2O$, and consider it related to graphitic acid. *Bulletin de la Société Chimique de Paris*, Vol. 23, No. 9.

BB.—Other Methods for determining Total Carbon.

A great number of methods have been devised for determining total carbon, some of which we will briefly outline, remarking, however, that the foregoing is entirely satisfactory.

1 *Method of* ALVARGONZALEZ. See *Am. Chem.*, Vol. V., p. 457.—Place 10 grms. of borings in a beaker and treat with a solution of cupric sulphate (40 grms. $CuSO_4$ in 200 200 c.c. H_2O), stirring until the reaction ceases. Add dilute HNO_3 gradually, and let stand until the copper has dissolved. Dilute the solution and filter through one of *Rother's half filters* (described in *Chem. News*, Jan. 30, 1874, p. 57), wash thoroughly, and dry on funnel at 100° C. Detach ppt. from filter carefully, place in a weighed crucible (throw away filter), dry at 100° C., and weigh. Ignite and weigh again; the difference between two weighings gives total carbon.

This method is not free from objections, but will answer when great accuracy is not indispensable, and speedy results are desirable.

2. *Method employed by* I. LOWTHIAN BELL. See *Chemical Phenomena of Iron Smelting*, London, 1872.—Digest 3 grms. borings from 24 to 48 hours with a solution of $CuSO_4$ in excess, collect the spongy Cu+C+graphite on an asbestos filter, and burn the carbon in a stream of oxygen gas, as in the ultimate analysis of organic bodies collecting the CO_2 in KHO solution. Cf. *Analysis No. 30.*

3. *Method of* REGNAULT *and* BROMEIS. See Crookes' *Select Methods*, p. 74.—Heat borings in a combustion tube with a mixture of plumbic chromate and $KClO_3$, collecting the CO_2 in KHO.

4. *Methods for the liberation of Combined Carbon* are also numerous.

(*a*) BOUSSINGAULT triturates the iron in a porcelain mortar with 15 to 20 parts of $HgCl_2$ and sufficient water to make a thin paste:

Then dilute with 200–250 c.c. HCl and warm for an hour; filter from the SiO_2+C, wash and dry. Transfer to a platinum boat, and heat in a current of pure H, volatilizing the Hg_2Cl_2. Weigh the C, heat again in a current of O, burning off the C, and weigh again.

(*b*) WEYL dissolves the pig-iron under the influence of a galvanic current. Attach a weighed piece of cast-iron to the positive pole of a Bunsen cell, and suspend it in dilute HCl. The iron dissolves, H being given off at the negative pole, and the carbon is separated.

Weyl has also devised another method based upon the following reaction:

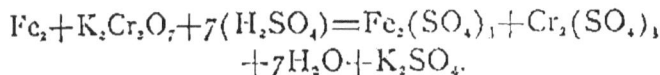

$$Fe_2+K_2Cr_2O_7+7(H_2SO_4)=Fe_2(SO_4)_1+Cr_2(SO_4)_3$$
$$+7H_2O+K_2SO_4.$$

See Crookes' *Select Methods*, p. 76.

(*c*) McCREATH'S *Method.* See *Engineering and Mining Journal*, March 17, 1877. The author uses double chloride of ammonium and copper to dissolve out the iron, while the precipitated copper dissolves in excess of this reagent; he then oxidizes the carbon by means of CrO_3 in an apparatus somewhat similar to Elliott's, collecting the CO_2 in a Liebig potash-bulb.

5. EGGERTZ *Colorimetric Method.* See Crookes' *Select Methods*, pp. 81 to 84; also Britton's paper in *Journal of the Franklin Institute*, May, 1870.

C. — Other Methods for the Determination of Sulphur and Phosphorus.

A. Dissolve 10 grms. KClO₃ in 200 c.c. H₂O, place in a 500 c.c. flask, add 5 grms. fine borings, boil and add 60 c.c. HCl, little by little, boiling until the Fe dissolves:

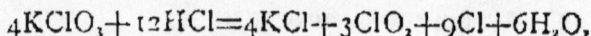

$$4KClO_3 + 12HCl = 4KCl + 3ClO_2 + 9Cl + 6H_2O,$$

and

$$2Fe + ClO_2 + Cl + 4HCl = Fe_2Cl_6 + 2H_2O.$$

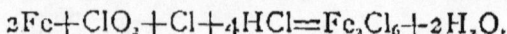

Evaporate, dry on water-bath to insure oxidation of sulphur. Thorough dryness is unnecessary, since SiO_2 does not interfere in acid solution with the precipitation of $BaSO_4$. Then add 10 c.c. $HCl + 30$ c.c. H_2O, and digest on water-bath until all the Fe_2Cl_6 is dissolved. Then add 20 c.c. H_2O, filter, and wash thoroughly. Add 2 c.c. of a saturated solution of $BaCl_2$ (enough to precipitate the H_2SO_4 from 0.1 grm. S); after cooling add 5 c.c. NH_4HO, stir and let stand 24 hours. Filter, and wash by decantation with cold water two or three times, and then thoroughly with hot water. Dry, ignite, and weigh. If the precipitate shows traces of iron after ignition, purify by solution in H_2SO_4.

B. For the determination of phosphorus dissolve the pig-iron in the same manner, and dry at 140° C; some anhydrous Fe_2O_3 will remain with the SiO_2; add water, filter, fuse residue with a little $KHSO_4$, soften with H_2SO_4, and dissolve in water. Filter from the SiO_2, and determine it as a check on the main analysis. Add filtrate to main one, and determine the P_2O_5 by means of ammonium molybdate, as in *Analysis* No. 21.

2. *Method of* DR. T. M. DROWN. See *Am. Chem.*, Vol. IV, p. 423.

Treat 5 grms. of borings in a flask with HCl, and pass the H_2S and PH_3 formed through a series of three bottles containing a solution of $K_2Mn_2O_8$ (1 grm. to 200 c.c. H_2O) Avoid a very rapid evolution of the gas; when this ceases, aspirate for some time, and then pour the contents of the bottle into a beaker, rinse with water, and add sufficient HCl to decompose the $K_2Mn_2O_8$. Filter the colorless so- lution, add $BaCl_2$, to throw down the H_2SO_4, and proceed as usual.

3. *Method employed by* J. Lowthian Bell.

Dissolve in HCl as above, and pass the gases through a solution of potassic plumbate (lead nitrate super-saturated with KHO). Boil half an hour, or until the evolution of gas has ceased. Wash the PbS formed, oxidize it with HNO_3, and throw down the S as $BaSO_4$ by means of $Ba(NO_3)_2$. Let stand 24 hours, collect on a filter, dry, ignite, and weigh. This method is said to give higher percentages of S than that of Eggertz. Compare *Fres.*, § 218, 3.

4. *Method of* Arthur H. Elliott. See *Am. Chem.*, Vol. I, page 376.

5. *Method employed by* Koninck and Dietz. See *Prac- tical Manual* of Chemical Analysis and Assaying applied to Iron. Translated by Robert Mallet. London, 1872.

Dissolve 3 to 5 grms. borings in HCl in a flask connected with four bottles, the first a condenser, the three following containing solution of $AgNO_3$ (1 part of nitrate to 20 parts of water). Boil, and when gas ceases to evolve, aspirate. Pour contents of flask on one filter, and wash the Ag_2S. Wash out the flask and cleanse the ends of the tubes with bromine water, and expel excess of Br by heat; the follow- ing reaction ensues :

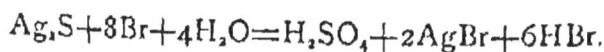

$$Ag_2S+8Br+4H_2O=H_2SO_4+2AgBr+6HBr.$$

The phosphide is also converted into phosphoric acid Filter from AgBr. and ppt. H_2SO_4 with $BaCl_2$ as usual.

6. *Method of* BOUSSINGAULT *for determination of Phos-phorus.* See *Annales de Chimie et de Physique,* June, 1875, and abstract in *American Chemist*, Vol. VI, p. 275.

7. For additional methods consult also papers by Alfred H. Allen, *Chem. News*, XXIX, p. 91, and paper by Hamilton, *Chem. News*, Vol. XXI, p. 147. Compare Crookes' *Select Methods*, pp. 84–89.

D.—Determination of Iron Manganese, etc.

The iron may be determined by difference or by Margue-rite's method, in which case dissolve 0.2 grms. of pig-iron in H_2SO_4, and proceed as usual. It is advisable to use a rather dilute solution of $K_2Mn_2O_3$ towards the close of the oxidation.

For the determination of the bases of *Groups* II, III, and IV, dissolve 10 or 20 grms of pig-iron in HCl, remove the SiO_2 by drying thoroughly, and proceed as in *Analysis No. 21.*

The manganese may be thrown down in the filtrate, from the basic acetate of iron by means of bromine, or in the absence of calcium, magnesium, etc., by hydrodisodic phos-phate. See *Fres.*, § 109, 3, also § 218, 6. For other methods of estimating manganese see articles by Samuel Peters in *Chem. News*, Vol. XXXIII, p. 35, and by William Gal-braith, in *Chem. News*, Vol, XXXIII, p. 47.

See also paper by Charles H. Piesse in *Chem. News*, Vol. XXIX., pp. 57 and 110.

For testimony as to the condition in which silicon exists in pig-iron, see paper by E. H. Morton, *Chem. News,* Vol. XXIX., p. 107.

ANALYSES NOS. 23 AND 24.—ARSENICAL NICKEL ORE.

May contain SiO_2, S, As, Sb, Pb, Cu, Fe, Al, Mn, Zn, Co, Ni, Cu, Mg, etc.

To be determined: As, Ni, Co.—Pulverize very finely; heat 2 grms. (or 4, if a very poor ore) in a covered casserole with fuming HNO_3 until the ore is completely dissolved, except a little silica. Expel excess of acid on a water-bath, add 10 c.c. HCl, dilute to about 200 c.c., warm and filter. Consult article by *Fresenius* in *Am. Chem.*, Vol. IV, p. 289.

Residue a.	Filtrate a.		
SiO_2, $PbSO_4$, $CaSO_4$, etc. Test with blowpipe for Co and Ni. If found, fuse with $KHSO_4$, and add water solution to *Filtrate a.*	Add a little Na_2SO_3 and heat, pass H_2S through the warm liquid until saturated. See *Fres.*, § 125, 1, and § 127, 4, *a.* Let stand some hours, throw on small filter, and wash with weak H_2S water.		
	Precipitate b.	*Solution b.*	
	As_2S_3. If much free S is mixed with the precipitate, dry, and exhaust with CS_2. Otherwise treat moist precipitate, filter and all, in a porcelain casserole with fuming HNO_3; expel excess of acid on water-bath, then dilute to about 150 c.c. and throw down As_2O_5 with "magnesia mixture." *Fres.*, § 127, *a.* Let stand 12 hours	Boil with a little $KClO_3 + 5$ c.c. conc. HCl, evaporate nearly to dryness, add water, warm, and add NH_4HO in excess. Wash well hot.	
		Precipitate c.	*Filtrate c.*
		$Fe_2O_3 \, 3H_2O$. Dissolve in HCl and reprecipitate with NH_4HO, filter and add the filtrate to *Filtrate c.*	Combine with second filtrate from the iron and add 20 c.c. NH_4HO. The solution should not measure more than 250 c.c. Introduce platinum electrodes and start the galvanic battery. (See *Am. Chem.*, Vol. VI, page 213.) Run battery all night, but take care to maintain an excess of NH_4HO. Wash the precipitated Co and Ni with water, alcohol, and weigh. To determine if precipitation is complete, test solution with NH_4HS.

		Precipitate d.	
in the cold. Filter through a weighed filter, collect filtrate and washings separately. Measure filtrate and test washings. Dry precipitate at 105° to 110° C., and weigh, repeat to a constant weight. For every 16 c.c. of filtrate (*not* washings) add one mgm. to the weight of the precipitate. See *Fres.*, § 92, c.	Test the precipitate of iron for nickel.	Dissolve in warm HCl. partly neutralize with KHO, add a conc. sol. of KNO_2 in excess, acidulate with acetic acid and let stand 24 hours. Filter and wash with a solution of neutral $KC_2H_3O_2$ (10 per cent. sol.), and afterwards with alcohol. See *Fres.*, § 111, 4.	**Filtrate d.** Boil down to smaller bulk, and test with electricity as above.
		Filtrate e. Concentrate, add KNO_2 and let stand. If precipitate forms add to precipitate e.	**Precipitate e.** Dissolve in HCl, neutralize with NH_4HO, and add 10 c.c. more NH_4HO. Throw down Co by the battery as before. Weigh. Weight of Co+Ni less weight of Co gives weight of Ni. Consult *Revue Universelle des Mines*, Vol. 32, p. 545.

Notes.—By the addition of sodium sulphite to *Filtrate a* we reduce the nitric acid and prepare the solution for the action of H_2S. The reaction is as follows:

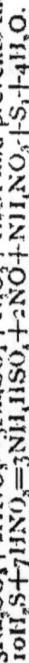

$$3Na_2SO_3+2HNO_3=3Na_2SO_4+N_2O_3+H_2O.$$ and prevents the action shown in this equation:
$$10H_2S+7HNO_3=3NH_4HSO_4+2NO+NH_4NO_3+S_7+H_2O.$$

Consult article by Parnell, in *Chem. News*, Vol. XXI, p. 133. On determination of cobalt and nickel consult article by E. Donath in *Chem. News.* Vol. XLI, p. 15.

Analysis No. 25.—GUANO.

Consult Fres., *Quant. Analysis,* § 220 ; also article by F.
A. Cairns, *Am. Chem.,* Vol. I, p. 82.

To be determined : SiO_2, CaO, MgO, Fe_2O_3, P_2O_5, SO_3,
H_2O, NH_3, total N, organic and volatile matter.

A.—Determination of Moisture.

Heat 1 grm. at 100° C. until constant weight and loss=
$H_2O [+(NH_4)_2CO_3]$.

In cases where great accuracy is required, a correction
for the $(NH_4)_2CO_3$ counted as water must here be made.
Heat the substance in a U tube in a water-bath and aspirate,
collecting the $(NH_4)_2CO_3$ in normal H_2SO_4. Titrate with
KHO as usual. Subtract $(NH_4)_2CO_3$ found from H_2O
$[+(NH_4)_2CO_3]$ determined by heating at 100° C. as above.

B.—Organic and Volatile Matter.

Determine loss by ignition in open crucible, and correct
for H_2O, (CO_2) and $(NH_4)_2CO_3$).

C.—Ammonia.

Use *Schlösing's* method, Fres., § 99, 3. b. Mix the guano
with milk of lime and place under bell-jar over a dish of
normal H_2SO_4. A large surface of acid in proportion to
the guano solution is desirable. Let stand, *cold,* 48 hours
or more, and titrate with normal KHO as in acidimetry.
(Cf. *Analysis No.* 11.)

D.—Total Nitrogen.

Use Varrentrapp and Will's Method, as detailed in Fres.,
§ 185. Heat the guano in a combustion tube with soda
lime, converting it into NH_3. Absorb the NH_3 in a standard solution of H_2SO_4, aspirate and disconnect bulb. Add
litmus and titrate with standard KHO.

E.—Sulphuric Acid.

Dissolve in hot HCl, filter and precipitate with $BaCl_2$, or
follow the Scheme F.

F.—Scheme for determination of SiO₂, Al₂O₃, Fe₂O₃, CaO, MgO, and total P₂O₅.

Dissolve 5 grms. of guano in HNO₃, evaporate to dryness, add HNO₃+H₂O, boil and filter.

Residue a.

SiO₂, and silicates. Wash, dry, ignite, and weigh, or fuse as in Analysis of dolomite.

Filtrate a.

Dilute filtrate and washings to 500 c.c. mix well, and divide into five unequal parts.

100 c. c.	100 c. c.	50 c. c.	50 c. c.	250 c. c.
Reserve for accidents.	Determine H_2SO_4 with $BaCl_2$ in the usual manner, if a duplicate be desired.	Determine the P_2O_5 with ammonium molybdate, as in *Solution g^1, Analysis No. 21.*	Determine the P_2O_5 in duplicate. Cf. *Note 12 of Analysis No. 21.*	Add 1 grm. iron wire dissolved in $HCl+HNO_3$. Precipitate iron, etc., as basic acetates, as in *Filtrate f, Analysis No. 21.* Wash the ppt., dissolve in HCl, dilute to 250 c. c., and divide into four unequal parts.

The 250 c. c. is divided into four unequal parts:

50 c. c.	50 c. c.	50 c. c.	100 c. c.
Reserve for accidents.	Determine total Fe by means of $K_2Mn_2O_8$, and deduct 20 per cent. for iron added.	Determine the Fe in duplicate. For details see *Analysis No. 3.*	Determine Fe in Al_2O_3 as in *Solution g^2, Analysis No. 21,* and in calculating allow for the iron added.

Analysis No. 26.—SUPERPHOSPHATE OF LIME.

To be determined: Moisture, reduced (or reverted) P_2O_5, soluble P_2O_5 and available P_2O_5.

A.—Determination of Moisture.

Dry 1 grm. at 100° and weigh—loss of weight=moisture.

B.—Determination of Total P_2O_5.

Weigh out 1 grm. accurately, mix with 2 grm. KNO_3 and 4 grms. Na_2CO_3, fuse in platinum crucible, dissolve in HNO_3, evaporate in a casserole to dryness (to dehydrate $SiO_2+aq.$), add water and filter. Wash thoroughly and dilute filtrate to 500 c.c.; take 50 c.c. of this solution (=0.1 grm. of superphosphate) and determine P_2O_5 with $(NH_4)_2 MoO_4$ as usual. Consult *Note* 12, *Analysis No. 21*.

C.—Determination of Insoluble P_2O_5.

Digest 1 grm. with about 400 c.c. of water in 8 to 10 different portions successively, rubbing the superphosphate with water in a porcelain mortar. Filter and treat residue (filter included) with about 50 c.c. solution of ammonium citrate containing 30 grm. of salt to 100 c.c. of water, and carefully neutralized if acid. Digest at about 70°C. for 40 minutes or longer, filter and wash. Dry residue and fuse exactly as in B. Estimate the P_2O_5 in same manner using 100 c.c. (=0.2 grm. of superphosphate) of the solution (500 c.c.) for each determination.

D.—Determination of Reduced and Insoluble P_2O_5.

Leach another sample (1 grm.) with water as in C (omitting the use of ammonium citrate), dry the residue and fuse as in B. Continue as in B, taking 150 c.c. of the solution for each determination.

E.—Calculation.

The reduced P_2O_5 is found by subtracting the P_2O_5 in C from that in D. The soluble $P_2O_5=$ B–D and the available $P_2O_5=$ B–C.

Note.—Reduced or reverted P_2O_5 forms thus:

$$Ca_3P_2O_8 + CaH_4P_2O_8 = 2Ca_2H_2P_2O_8.$$

Consult *Bolley's Handbuch*, pages 802–806.

Analysis No. 27.—POTABLE WATER.

To be determined: K; Na; Mg; Ca; Cl; SO_3; SiO_2; organic and volatile matter, total solids; hardness (soap test); oxygen required to oxidize organic matter (permanganate test).

Quantity required, three to four gallons; collect in clean demijohns.

A.—Determination of Total Solids and Loss by Ignition.

Measure out 250 c.c. of the water, evaporate to dryness on a water bath, in a weighed platinum dish of 100 c.c. capacity. During the evaporation cover the dish with a paper screen. Dry in an air bath 120°–130° C. and weigh. Weight of residue gives "Total Solids." Ignite gently over a Bunsen burner, moisten with a solution of CO_2 in distilled H_2O, dry on water-bath, heat in air-bath 120°–130° C. as before and weigh; difference between second and first weights gives organic and volatile matter, also called "Loss by Ignition." For further treatment of residue see *F*. Compare Chapter II of Wanklyn's " *Water Analysis*," 3rd edition, 1874.

B.—Determination of SiO$_2$, Fe$_2$O$_3$, Al$_2$O$_3$, CaO, MgO.

Evaporate 4 to 6 litres of water (according to the proportion of total solids) to small bulk in porcelain dish. Add HCl, transfer to platinum dish, washing carefully the porcelain dish, evaporate to dryness, filter from SiO$_2$ and follow scheme for Dolomite, *Analysis No. 7.*

C.—Determination of H$_2$SO$_4$.

Take 1 litre (or less) of the water, boil down to 200 c.c. with a few drops of HCl, and determine SO$_4$ as BaSO$_4$ in the usual manner. If the water contains sufficient H$_2$SO$_4$ (as sulphates) to give a feeble precipitate with BaCl$_2$ before concentration, one-half or one-quarter litre will suffice.

D.—Determination of Cl.

Test the water with AgNO$_3$ for Cl, and if no cloud is formed evaporate 1 litre to small bulk; otherwise 25 to 50 c.c. suffice. Add a slightly acid solution of AgNO$_3$, and proceed as usual.

Second method. Determine the Cl volumetrically by a standard solution of AgNO$_3$, using potassium chromate as an indicator. See Fres. § 141,I, *b, a.*

E.—Determination of Na and K.

Evaporate 6 litres to dryness in a large porcelain dish, finishing on a water bath. Boil the residue with distilled water several times, filter into a platinum dish and wash. Add Ba(HO)$_2$ to filtrate. Evaporate to dryness, heat to low redness, let cool, take up with water, add (NH$_4$)$_2$CO$_3$ and a little (NH$_4$)C$_2$O$_4$, wash the precipitate, filter, add HCl to filtrate, evaporate to dryness, ignite and weigh. Dissolve in water and if not clear, filter, evaporate, dry, ignite cautiously, and weigh again. This residue of NaCl+KCl must be perfectly white and soluble without residue in water. Determine the Cl in the weighed NaCl+KCl and calculate the Na and K as in Fres. page 841. Compare *Wanklyn, 3rd edition,* page 63.

F.—To check determination of Na and K.

Moisten the weighed "Total Solids" of A with dilute H_2SO_4, dry and ignite with a little powdered $(NH_4)_2CO_3$ to constant weight. By deducting from this weight, *calculated for one gallon,* the combined weights of SiO_2, Fe_2O_3, Al_2O_3, CaO, MgO (the latter four reckoned as sulphates), the weights of Na_2SO_4 and K_2SO_4 are obtained.

G.—Dr. Clark's Soap Test.

Consult Sutton's *Volumetric Analysis,* §83,10: or Wanklyn's *Water Analysis,* 3rd edition, page 125.

Principle.—Hard water, so called, destroys much soap before a lather is formed, owing to the formation of insoluble salts, viz.: stearates, palmitates, and oleates of calcium and magnesium.

Preparation of Soap Solutions.—Dissolve 10 grms. of good white Castile soap (which should contain about 12 per cent. of water) in 1 litre of alcohol 90 to 95 per cent. Let stand and siphon off from the residue. Label this solution "No. 1." Take of solution No. 1. 100 c.c.; of 56 per cent. alcohol, 65 c.c.; of distilled water, 75 c.c., and mix. Label this soap solution "No 2."

Preparation of Standard Calcium Solution.—Dissolve 1 grm. of precipitated $CaCO_3$ in HCl, evaporate until neutral, take up with water and dilute to 1000 c.c. 1 c.c. of this solution contains grm. 0.001 $CaCO_3$.

Standardization of Soap Solution.—Fill a burette with soap solution No. 2. Place 10 c.c. of calcium solution in a glass stoppered bottle. Add it to 100 c.c. of distilled water, run in soap solution from the burette and shake well, and continue adding soap solution until a lather is formed of sufficient consistence to remain for five minutes on the surface of the water. Read burette and calculate. Repeat.

In certain cases allowance should be made for the amount of soap solution destroyed by water itself; 100 c.c. destroys 0.8 c.c. soap solution.

Performance of Analysis.—Same as above. Report milligrammes per litre and grains per gallon of $CaCO_3$.

Example.—10 c.c. of the standard solution of chloride of calcium required 23 c.c. of soap solution—

But 10 c.c. of $CaCl_2$ solution is equivalent to .01 grm. of $CaCO_3$, hence

$$\left.\begin{array}{c} \text{c.c. used} \\ 23 \end{array}\right\} \ 1 \ \text{c.c.} \ \left.\begin{array}{c} \\ \end{array}\right\} = \left.\begin{array}{c} .01 \\ \text{grm. } CaCO_3 \end{array}\right\} : a = .00043 \ \text{grms.}$$

And if 100 c.c. of water under examination require 33 c.c. of $CaCl_2$ solution, we have $33 \times .00043 \times 10 =$ grms. per litre of $CaCO_3$. This gives .1419; and $.1419 \times .058318$ gives grains per gallon. For the factor .058318 consult I, *Calculation of Results.*

H. Permanganate Test for Organic Matter.

Principle.—Permanganate of potassium in solution oxidizes putrescible organic matter.

Preparation of solution of permanganate. — Dissolve 0.320 grms. of permanganate in 1 litre of water. Dissolve 0.7875 pure oxalic acid in 1 litre of water, weighing very accurately. Of this solution 1 c.c.=0.0001 grm. oxy gen. To standardize the permanganate, take 10 c.c. of oxalic acid solution; dilute to 100 c.c. with distilled water, add 5 c.c. dilute H_2SO_4, heat nearly to boiling and run in from a burette the permanganate solution. 10 c.c. of oxalic acid will require 12 to 15 c.c. permanganate. Calculate value of 1 c.c. of latter in milligrammes of oxygen.

Testing water.—Take 100 c.c. potable water add H_2SO_4, add standardized permanganate, little by little in the cold, until the water retains a pink tinges after one-half hour's standing. Report amount of oxygen required to oxidize organic matter.

Example.—10 c.c. of standard solution of oxalic acid required 14 c.c. of solution of $K_2Mn_2O_8$—

But 10 c.c. of $H_2C_2O_4$ solution is equivalent to 0.001 grms. of oxygen, hence

$$\left.\begin{matrix} \text{c.c. used} \\ 14. \end{matrix}\right\} : 1. \text{ c.c.} \left.\right\} = \left.\begin{matrix} 1. \text{ mgm.} \\ \text{oxygen} \end{matrix}\right\} : a = .0713 \text{ mgms.}$$

And if 100 c.c. of water under examination required 0.8 c.c. of $K_2Mn_2O_8$, we have $0.8 \times .0713 \times 10 =$ mgms. per litre of oxygen required to oxidize organic matter. This gives .5704 milligrammes and $.5704 \times .058318$ gives grains of oxygen per gallon. See I, *Calculation of Results.*

I.—Calculation of Results.

To convert *grms. in a litre* into *grains in a gallon*, multiply the number of milligrammes of each constituent by 0.058318; or use Dr. Waller's Table, published in *Am. Chem.*, Vol. V, p. 278. Report results in two ways: the grains per gallon of uncombined constituents, viz., SiO_2, Fe_2O_3, Al_2O_3, CaO, MgO, Na_2O, K_2O, Cl, SO_3, together with "Loss by Ignition" and "Total Solids;" and secondly report the grains per gallon of the bases combined with acids in accordance with the following scheme.

Combine   K   as K_2SO_4
"   excess of   K   "   KCl
"   "   "   Cl   "   Na Cl
"   "   "   Na   " Na_2SO_4
"   "   "   Cl   " $MgCl_2$
"   "   " SO_4 " $CaSO_4$
"   "   "   Ca   " $CaCO_3$
"   "   "   Mg   " $MgCO_2$.

The sum of the combined salts + "Loss by Ignition" should equal the "Total Solids" very nearly.

Example, showing method of calculation.—A sample of potable water yielded on analysis the following results:

Cl   .215 grains per gallon.
Na   .291   "   "   "
SO_3 .340   "   "   "
CaO   .804   "   "   "
etc.   etc.

Begin by calculating the amount of Na required to saturate the Cl found, thus:

$$(1) \begin{cases} Cl : Na = \left\{ \begin{array}{l} Amount \\ of\ Cl\ found. \end{array} \right\} : \left\{ \begin{array}{l} Amount\ of\ Na \\ needed\ for\ the\ Cl. \end{array} \right\} \\ 35.5 : 23 = 0.215 \qquad : \quad w \end{cases}$$

$w = 0.139$ grains.

hence $0.215 + 0.139 = 0.354$ grains NaCl.

But the water contains .291 grains Na, hence we have $.291 - .139 = .152$ grains Na left over to combine with SO_3.

0.152 grains Na corresponds however to 0.204 grains Na_2O making then a proportion similar to (1) we have

$$(2) \begin{cases} Na_2O : SO_3 = \begin{cases} \text{Amount} \\ \text{of } Na_2O \\ \text{remaining.} \end{cases} : \begin{cases} \text{Amount of } SO_3 \\ \text{needed} \\ \text{for the } Na_2O. \end{cases} \\ 62 \quad : 80 = 0.204 \quad : \quad x \end{cases}$$

$$x = 0.263 \text{ grains.}$$

hence $0.204 + 0.263 = 0.467$ grains Na_2SO_4.

But the water contains 0.340 grains SO_3 hence we have $0.340 - 0.263 = 0.077$ grains SO_3 left over to combine with CaO.

Accordingly we have the proportion

$$(3) \begin{cases} SO_3 : CaO = \begin{cases} \text{Amount of } SO_3 \\ \text{remaining.} \end{cases} : \begin{cases} \text{Amount of CaO} \\ \text{needed for the } SO_3. \end{cases} \\ 80 \quad : \quad 56 = 0.077 \quad : \quad y \end{cases}$$

$$y = 0.0539 \text{ grains CaO,}$$

hence $0.077 + 0.0539 = 0.130$ grains $CaSO_4$.

Proceeding in like manner the CaO remaining is regarded as combined with CO_2.

$0.804 - 0.0539 = 0.7501$ grains CaO; and since.

$$(4) \begin{cases} CaO \quad : \quad CaCO_3 = 0.7501 \quad : \quad z \\ \text{whence} \quad z = 1.34 \text{ grains } CaCO_3. \end{cases}$$

Collecting the results of the calculation we have (thus far) the following figures for the *constituents combined*:

$$NaCl = 0.354 \text{ grains per gallon.}$$
$$Na_2SO_4 = 0.467 \quad " \quad " \quad "$$
$$CaSO_4 = 0.130 \quad " \quad " \quad "$$
$$CaCO_3 = 1.34 \quad " \quad " \quad "$$
$$\text{etc.,} \quad \text{etc.}$$

The following will serve as a further example of the manner of reporting similar analyses.

ANALYSIS OF CROTON WATER BY DR. C. F. CHANDLER.

	Grains per gallon.
Soda	0.326
Potassa	0.097
Lime	0.983
Magnesia	0.524
Chlorine	0.243
Sulphuric acid	0.322
Silica	0.621
Alumina and oxide of iron	trace
Carbonic acid (calculated)	2.604
Water in bicarbonates (calculated)	0.532
Organic and volatile matter	0.670
	6.927
Less oxygen equivalent to the chlorine	.054
Total	6.873

These acids and bases are probably combined as follows:

	Grains per gallon.
Chloride of sodium	0.402
Sulphate of potassa	0.179
Sulphate of soda	0.260
Sulphate of lime	0.158
Bicarbonate of lime	2.670
Bicarbonate of magnesia	1.913
Silica	0.621
Alumina and oxide of iron	trace
Organic matter	0.670
	6.873

Analyses No. 28 and No. 29. — SPECIFIC GRAVITIES OF
SOLIDS AND LIQUIDS.

A—Sp. gr. of a solid by direct weight.

Weight of solid in the air $= w$

"   "   "   " water $= w'$

$$Sp. gr. = \frac{w}{w - w'}$$

B.—Sp. gr. of a solid by the flask.

Weight of solid $= w$

"   " flask $+$ water $= w^{\text{I}}$

"   "   "   "   " $+$ solid $= w''$

$$Sp. gr. = \frac{w}{(w + w') - w''}$$

C.—Sp. gr. of a body soluble in water.

Weight of body in air $= w$

"   "   "   " oil $= w^{\text{I}}$

Sp. gr. of oil $= a$

"   "   " water $= \text{I}$

The liquid displaced being

$$w - w^{\text{I}} = w''$$

then

$$a : \text{I} = w'' : w'''$$

$$Sp. gr. = \frac{w}{w'''}$$

D.—Sp. gr. of a body lighter than water and insoluble in it, e.g., Cork.

$$\text{Sp. gr.} = \frac{w}{w'-w''+w}$$

E. — Sp. gr. of a Body lighter than Water and soluble in it.

$$\begin{aligned}
\text{Weight of body in air} &= w \\
\text{"\quad"\quad"\quad" naphtha} &= w' \\
w - w' &= w''
\end{aligned}$$

$$\begin{aligned}
\text{Sp. gr. of naphtha} &= A \\
\text{"\quad"\quad" water} &= I
\end{aligned}$$

$$A : w'' = I : w'''$$

$$\text{Sp. gr.} = \frac{w}{w'''}$$

F. — Determination of the Proportion of two Metals in an Alloy.

$$\begin{aligned}
\text{Sp. gr. of the alloy} &= S \\
\text{Weight of the alloy} &= A \\
\text{Sp. gr. of one of the metals} &= s' \\
\text{Sp. gr. of the second metal} &= s'' \\
\text{Weight of one metal} &= w' \\
\text{Weight of the second metal} &= w''
\end{aligned}$$

$$w' = A \frac{(S'-s'')s'}{(s'-s'')S}$$

$$w'' = A - w'$$

For proofs of this formula, see *Galloway's First Step in Chemistry*, p. 74.

G. — Sp. gr. of a liquid by the flask.

$$\begin{aligned}
\text{Weight of flask} &= F \\
\text{"\quad"\quad" and water} &= w \\
\text{"\quad"\quad"\quad" liquid} &= w'
\end{aligned}$$

$$Sp.\ gr. = \frac{w' - F}{w - F}$$

H. — Sp. gr. of a Liquid by weighing a Substance in it.

Weight of substance   = w
"   "   " in liquid   = w'
Sp. gr. of the substance   = A

$$w : (w - w') = A : sp.\ gr.$$

$$or\quad Sp.\ gr. = \frac{(w - w')\,A}{w}$$

Analysis No. 30, 31, *and* 32.   *Organic Analysis.*

INTRODUCTORY NOTES. The analysis of organic bodies comprises two branches ; PROXIMATE ANALYSIS which deals with the separation of *proximate principles* of organic bodies without altering them, and ULTIMATE ANALYSIS, by which the nature and quantity of the *elements* composing the organic bodies are determined.

No systematic course of proximate analysis is possible in the present state of the science; animal chemistry is in this respect more advanced than vegetable ; for a course of zoo-chemical analysis see article by *Gorup-Besancz* in the *Neues Handwörterbuch der Chemie*, I, 551, and compare *Watt's Dictionary*, I, 249. See also *Heintz Lehrbuch der Zoochemie* and *Lehman's Physiological Chemistry.* For general principles of proximate organic analysis, consult *Dr. Albert B. Prescott's " Outlines of Proximate Organic Analysis,"* a most useful manual, and the only one of its kind. For special methods of analyzing organic bodies, especially of commercial articles, consult *" Bolley's Handbuch der Technisch-chemischen Untersuchungen,"* of which the second edition by Emil Kopp is most valuable.

The method of conducting an ultimate analysis is sufficiently detailed in *Fresenius' System*, § 171-189, yet the following summary may be of service in calling attention to the chief points.

A. Determination of C, H, and O, in Sugar.

Select a very pure well crystallized sample of sugar, rock-candy will do, but small crystals from a vacuum pan are better. Dry at 100° C, in powder.

Provide the following articles : —

(1) The dried substance in a tared watch glass.

(2) Combustion tube of hard glass drawn out as shown in *Fres.* § 174, cleaned and carefully dried.

(3) Liebig potash bulb filled with a KHO solution of Sp. gr. 1.27, or a U-tube filled with soda-lime.

(4) Chloride of Calcium tube ; that of the form described by *Thorpe* in his *Quant. Chem. Analysis* page 347, fig. 80 is advantageous.

(5) Small U-tube containing potash-pumice in one limb and $CaCl_2$ in the other.

(6) Rubber tubing.

(7) Fine wire for binding the tubing.

(8) Good corks, free from holes, rolled and pressed.

(9) Cupric oxide, granulated preferred, chemically pure, freshly ignited to remove organic matter and moisture, and contained in a corked holder.

(10) A platinum boat to contain the substance, or if another process be followed, a mixing wire.

(11) Combustion furnace.

(12) If oxygen is to be employed, a cylinder of this gas and a system of drying U-tubes must be provided.

(13) Sundry articles, such as glazed paper, agate mortar, towel, asbestus, a ramrod for cleaning combustion tube. etc.

Process of the Combustion.

(*a*) Weigh the substance (sugar) and preserve in a desiccator until ready for use ; weigh also the KHO bulb together with the U-tube (5), CaCl₂ tube.

(*b*) Dry the combustion tube and fill with cupric oxide ; the substance may be inserted on a platinum boat if the combustion is to be conducted with oxygen, otherwise it must be intimately mixed with some powdered CuO in the agate mortar and transferred by the glazed paper to the combustion tube. Stir also with the iron mixer. Avoid introducing moisture.

(*c*) Connect the apparatus, arranging it as shown in the cut on page 433 of *Fresenius' System.* Test the joints by heating the air in that bulb of the KHO apparatus which is between the solution and the combustion tube ; drive out a few bubbles of air and let cool, if an unequal level of the solution is maintained, the joints are tight.

(*d*) Conduct the ignition, heating gradually, and beginning at the end next to the CaCl₂ tube ; do not apply heat to the substance until several inches of CuO are red hot. Pass oxygen gas through the tube if that method is employed. *Fres.* § 178. The combustion of sugar may be completed in about half an hour, other substances require more time, especially those rich in Carbon.

(*e*) Aspirate air, or pass oxygen through the apparatus slowly.

(*f*) Disconnect the weighed tubes, cool and weigh. From the CO_2 and the H_2O found, calculate the C and the H respectively. The O is found by difference.

Theoretical Composition of Cane Sugar.

$$C_{12} \quad 144 \quad . \quad . \quad 42.11$$
$$H_{22} \quad 22 \quad . \quad . \quad 6.43$$
$$O_{11} \quad 176 \quad . \quad . \quad . \quad 51.46$$

$$\overline{342} \qquad 100.00$$

In the case of nitrogenous bodies introduce copper turnings or a spiral of sheet copper in the end of the combustion tube next to the absorption tubes ; the metallic copper at a red heat reduces any nitric oxide which may form, and the inert nitrogen passes through the absorption tubes without increasing their weight. See *Fres.* § 183.2.

The difficulty of effecting a complete oxidation of the carbon in organic substances increases, other things being equal, with the percentage of carbon contained in the substance ; the richer the substance in carbon, the smaller the amount should be taken for combustion. Moreover, it is desirable to graduate the quantity used, to prevent the formation of too large a quantity of carbonic anhydride to admit of complete absorption by the potash solution ; hence the following Table, used in Prof. A. W. Hoffman's Laboratory, University of Berlin, is of service in determining the amount of substance which may be conveniently employed.

Table showing amount of Substances to be used in Ultimate Analysis.

Of substances containing 80 percent carbon take 0.200 grms.

"	"	75	"	"	"	0.225	"
"	"	70	"	"	"	0.250	"
"	"	65	"	"	"	0.275	"
"	"	60	"	"	"	0.300	"
"	"	55	"	"	"	0.325	"

Of substances containing 50 percent carbon take 0.350 grms.

"	"	45	"	"	"	0.375	"
"	"	40	"	"	"	0.400	"
"	"	35	"	"	"	0.425	"
"	"	30	"	"	"	0.450	"
"	"	25	"	"	"	0.475	"
"	"	20	"	"	"	0.500	"

C. — Determination of Nitrogen in Potassium Ferrocyanide by Conversion into Ammonia.

Method of Varrentrapp & Will. See *Fres.* § 185.

Purify about 50 grms. of the commercial salt by recrystallization ; dry the crystals on filter paper and preserve in a desiccator. The crystallized salt contains 3 molecules of water.

Principle : When organic substances are heated with hydroxides of the alkaline metals the carbon is oxidized by the oxygen of the hydroxide, and hydrogen is set free ; if, however, nitrogen is present it combines with the nascent hydrogen, forming ammonia. (For an exception, see **D.**) By conducting the operation in such a way as to complete the reaction, and collecting all the ammonia by absorption in acid of known strength, the amount of nitrogen is easily calculated.

Requisites : The apparatus needed is, in general, the same as that used in determination of C and of H, but a somewhat shorter tube (40 cm.) may be used ; the ammonia is absorbed by normal sulphuric acid placed in pear-shaped bulbs of the form shown in Fig. 92, or Fig. 94, pages 443 and 445 of *Fres. System.* The substance used to oxidize the carbon is soda-lime, at present a commercial

article; it should be heated in a porcelain dish to expel
water and ammonia before using.

Operation: Fill the combustion tube about one-third full
of warm soda-lime and let it cool; then mix this in an
agate mortar with 0.2 to 0.4 grms. of the dry ferrocyanide
of potassium, and introduce the mixture again into the
tube; rinse the mortar with a little soda-lime, and then
fill the tube with the same nearly to the open end. Insert
a small plug of asbestos loosely, attach the absorption bulb
containing the sulphuric acid by a well-fitting cork, and
place the tube in the combustion furnace. Begin to heat
the tube at the end nearest the cork, and proceed gradually towards the other end.

The gas evolved should bubble quietly through the absorption tube, and when it ceases to pass break the tail-
piece of the combustion tube, and aspirate gently through
the whole apparatus.

Detach the absorption tube, empty its contents into a
beaker, rinse well, add a little litmus, or cochineal solution, and determine, by means of normal KHO, the
amount of acid remaining unneutralized by the ammonia.
For details of this process see *Analysis No.* 12.

Theoretical Composition of Potassium Ferrocyanide:

C_6 . . . . 17.1
N_6 . . . . 19.9
Fe 13.3
K_4 . . . . 37.0
$3H_2O$ 12.7
                                                    ———
                                                  100.0

D. — Determination of N from the Volume.

Dumas' method modified by Melsens, Cf. *Fres.* § 184. See also Watts' Dictionary, I. 242.

When nitrogen exists in an organic substance in the form of an oxide, e. g. nitro-benzol $C_6H_5(NO_2)$, Varrentrapp & Will's method cannot be employed because the oxides of nitrogen are not completely converted into ammonia on heating with soda lime. Dumas' method consists in heating the substance with oxide of copper, and measuring the nitrogen evolved by collecting over mercury. The process originally devised by Dumas necessitated the use of an air-pump to exhaust the combustion tube, but this may be obviated by following Melsens, who introduces hydro-sodium carbonate into the tube which gives up carbonic anhydride on heating, and drives out the nitrogen before it.

For Melsen's process provide the following articles :

(1) A combustion tube 70 cm. long.

(2) Mercury trough.

(3) Graduated cylinder.

(4) Copper oxide.

(5) Solution of potassium hydrate.

(6) Hydrosodium carbonate.

(7) Connecting tube.

(8) Corks, asbestos, rubber tubing, etc.

(9) Combustion furnace.

In filling the combustion tube observe the following order : Insert, first, 15 cm. of hydrosodium carbonate, then 5 cm. of copper oxide, then 15 cm. of copper oxide mixed with the substance to be analyzed, next add about 28 cm. of copper oxide, insert a copper spiral 5 cm. long, and lastly a plug of asbestos in the remaining 2 cm. Insert cork with connecting tube, and arrange apparatus as shown in Fig. 93. page 635, of *Fres. System.*

Conduct the operation as follows : Heat a portion of the NaHCO$_3$ until all the air is expelled ; test with a solution of KHO in an inverted test-tube ; then heat CuO to redness, arrange the graduated cylinder containing KHO solution over mercury, and heat the mixed CuO and substance until gas ceases to come off; lastly, expel the nitrogen in the combustion tube by again heating the NaHCO$_3$, some of which must have been left undecomposed. (Oxalic acid may be substituted for the HNaCO$_3$. See *Thorpe*, page 332.) Transfer the graduated cylinder to a vessel of water, hold it so that the level of the water within the cylinder and without is equal, then read off the volume of the gas in cubic centimeters, and simultaneously the temperature of the water and the height of the barometer.

Calculation of Results. To obtain the weight of nitrogen from its volume employ the following formula :

Let V = Volume of N observed, expressed in cubic centimeters.

And t° = Temperature of the gas.

" B = Height of the barometer expressed in millimeters.

" f = Tension of aqueous vapor at the temperature t°, expressed in mm. of mercury.

Then if W = weight of nitrogen we have :

$$ W = .0012566 \text{ V} \frac{1}{1+.00367 t°} \frac{B-f}{760} $$

The constant 0.0012566 is the weight in grammes of 1 c. c. of N at 0° C and 760 mm. The constant 0.00367 is the coefficient of expansion of gas.

Example : In an analysis of Butyramide —

C$_4$H$_7$O ⎫
H　⎬ N, the following data were obtained :
H　⎭

0.315 grms. of substance gave 43.9 c. c. N at $t°=17°3$ C and $B = 753.2$ mm.

First look out in a table the value of f at $17°.3$. (*Fres.*, page 837, Table.) We find (calculating for the tenths of a degree) $f = 14.7$.

Now V = 43.9 c. c.

$B — f = 753.2$ mm. $— 14.7 = 738.5$ mm.

And $1 + .00367 \times t° = 1.0635$.

Substituting in equation :

$$W = .0012566 \text{ V} \frac{1}{1+.00367t°} \frac{B—f}{760} \text{ we have :}$$

$$W = \frac{.0012566 \times 43.9 \times 738.5}{1.0635 \times 760} = 0.0504 \text{ grms. N.}$$

And $\frac{0.0504 \times 100}{0.315} = 16.00$ per cent nitrogen.

Theoretical Composition of Butyramide :

C_4 .	55.2
H_9	10.3
O' .	18.4
N .	16.1
	100.0

Analysis No. 33. — URINE.

For brief methods of analysis consult Dr. George B. Fowler's " Urine Analysis," Thudicum's " Manual of Chemical Physiology," pages 178–192, and Sutton's " Systematic Handbook of Volumetric Analysis," part vi. § 78. For figures of sedimentary deposits examine Ultzmann & Hofmann's " Atlas der Physiologischen und Pathologischen Harnsedimente." (44 plates.)

The following works may also be studied : Legg's " Guide to the Examination of Urine," Attfield's " Chem-

istry," F. Hoppe-Seyler's "Handbuch der Physiol. and
Pathol. Chem. Analyse," Neubauer & Vogel's "Anleitung
zur Qualitative und Quantitative Analyse des Harns,"
Gorup Besanez' "Lehrbuch der Physiologischen Chemie,"
pages 576–580, Ultzmann & Hofmann's "Anleitung zur
Untersuchung des Harns."

Constituents of Urine.

Urine, the secretion of the kidneys, in a healthy individ-
ual, is a clear, yellowish, fluorescent liquid of a peculiar
odor, saline taste, with a mean sp. gr. 1.020. The following are its normal constituents :

1. *Water.* — H_2O.

2. *Inorganic Salts.* — K, Na, NH, Ca, Mg, combined with
   HCl, H_3PO_4, H_2SO_4, CO_2, (HNO_3,) and SiO_2.

3. *Nitrogenous crystalline bodies.* — Urea, uric acid, hippuric acid, creatine, creatinine, xanthine, (ammonia,)
   cystine.

4. *Non-nitrogenous organic bodies.* — Sugar, lactic, succinic,
   oxalic, formic, malic, and phenylic acids, all in small
   quantities.

5. *Pigments.* — Urochrome, urohaematin.

6. *Albumenoid matters.*

7. *Matters derived directly from the food.*
   Besides these, urine may contain, under varying circumstances, as in disease, a large number of

8. *Abnormal constituents.* — Blood, pus, mucus, albumen,
   fibrin, casein, fats, cholesterin, leucine, tyrosine, allan-
   toin, taurine, biliary pigments, indigo-blue, melanin,
   glucose, inosite, acetone, butyric acid, benzoic acid,
   oxaluric acid, taurocholic acid, glycocholic acid, and
   many others. (See Watts' Dictionary, vol. v. p. 962.)

These substances do not occur simultaneously in all urine, and many of them but rarely. Only those commonly determined are considered in the Scheme (page 112).

Chemical Composition of Urine. (DALTON.)

Healthy. — Numbers Approximate.

Water 938.00
Urea 30.00
Creatine 1.25
Creatinine 1.50
Urate of soda ⎫
" potassia ⎬ 1.80
" ammonia ⎭
Coloring matter and mucus 30
Bi-phosphate of soda ⎫
Phosphate of soda ⎪
" potassa ⎬ 12.45
" magnesia ⎪
" lime ⎭
Chlorides of sodium and potassium 7.80
Sulphates of soda and potassa 6.90
———————
1000.00

Morbid urine may contain, also:

Albumen, (Bright's disease.)
Sugar, (Diabetes.)
Bile,
Excess of Urea,
Oxalate of calcium.

Action of Reagents on Urine.

Boiling acid urine effects no change.

Boiling alkaline urine makes it turbid if rich in earthy phosphates.

HNO_3 or HCl darkens the color, and throws down uric acid on standing.

KHO or NH_4HO throws down earthy phosphates.

$BaCl_2$ or PbA, in acidified urine, yield a white ppt. of sulphates. ·

$AgNO_3$ white ppt. of chlorides, also coloring matter and some organic substances.

Murexid Test. — Collect some of the uric acid thrown down by HCl, remove supernatant liquid, add conc. HNO_3, and evaporate to dryness. When cold add a drop of NH_4-HO. A purplish-crimson color shows formation of murexid ($C_8H N_6O_6$).

Reactions of Urea. — $Hg (NO_3)_2$ throws down a gelatinous white ppt. containing $COH_4N_2 .2HgO$.

Boiling with KHO converted into NH_4HO; test with Nessler reagent.

HNO_3, nitrate of urea precipitates.

$NaClO$ or $NaBrO$ decomposes urea with evolution of N.

Scheme for Analysis of Urine.

1. PHYSICAL CHARACTERS.

(a) *Odor.* — Certain peculiarities in odor indicate either nature of food or symptoms of disease.

(b) *Consistence.* —Viscous or fluid.

(c) *Color.*—When healthy, urine is amber-colored; when bilious, brown or greenish.

(d) *Specific Gravity.* — By the urinometer, 1015 to 1025 is marked H. S., signifying Healthy State. 4° c. makes a difference of about 1° in the reading.

2. TEST WITH LITMUS PAPER, and note whether acid or alkaline.

3. POUR A SAMPLE into a stop-cock funnel, and let stand 12 hours. If a deposit forms, filter, and examine the filtrate and sediment separately. Filtered urine leaves a scum of mucus. (For sediments, see Schemes, page 117 and 118.)

4. DETERMINE TOTAL SOLIDS. Evaporate 4 to 6 c. c., weighed, to dryness in a weighed dish. Dry at 115 c. (Inaccurate).

5. ASH. Evaporate 100 c. c. urine and ignite residue.

6. DETERMINATION OF UREA. CH_4N_2O.

A. — Liebig's Method.

Principle : Mercuric nitrate added to a solution of urea gives a white, gelatinous ppt. containing 1 molecule urea, and $2HgO$. (Absence of $NaCl$ necessary.)

Requirements :

(*a*) Standard solution $Hg (NO_3)_2$.

(*b*) Baryta solution.

(*c*) Carbonate of soda test paper.

(*a*) Standard solution of mercuric nitrate. Dissolve 72 grms. pure dry HgO in strong HNO_3, (50 grms.,) evaporate until syrupy, and dilute to 1 litre. If a yellow ppt. is produced by dilution, too little acid is present. It must be evaporated down, fresh acid added, and again diluted. 1 c. c. $= 0.01$ grm. urea. To test the strength of the mercuric nitrate dissolve 2 grms. cryst. urea in 100 c. c. water. 1 c. c. mercuric solution should equal 0.01 grm. urea.

(*b*) Solution of $Ba(NO_3)_2 + BaH_2O_2$. Mix 1 part cold saturated solution $Ba(NO_3)_2$ with 2 parts cold saturated solution BaH_2O_2, and add 3 parts distilled water.

(*c*) Soda test paper. Dip a sheet white filter paper into conc. sol. $Na_2(CO_3)$ and dry.

Process: Collect the urine passed during 24 hours, and measure carefully. Place 20 c. c. in a small beaker, add 20 c. c. barium solution, filter from the sulphates and phosphates. Of the filtrate 20 c. c. (containing 10 c. c. urine) are measured off, a drop of $AgNO_3$ added to precipitate excess of chlorides, and then standard solution of mercuric nitrate is added until a drop of the mixed solutions gives a yellow stain (of mercuric hydrate) on the test paper.

Byasson adds some of a solution of KHO (25 grms. to 1 litre water) from time to time to partly neutralize the acid set free. The solution must not be rendered alkaline.

Calculation: Amount urine passed in 24 hours $= A$; c. c. mercuric solution used $= C$; each c. c. being equal to 0.01 grm. urea; then $\dfrac{A \times C}{10} = $ grms. urea passed in 24 hours.

Caution: The urine must be free from phosphoric and hippuric acids. Consult Caldwell's "Agricultural Analysis," page 220. Urine must contain 2 per cent. urea. Cf. Watts' Dict. vol. v. p. 967.

B. — *Davey's Method of Estimating Urea.*

Pour a small quantity of urine into a graduated glass tube one-third full of mercury. Fill the tube with a solution of sodic hypochlorite, close tube, and invert quickly over a saturated solution of NaCl. Let stand several hours while the following reaction ensues :

$$CH_4N_2O + 3(NaClO) = CO_2 + 2H_2O + 3NaCl + N_2$$

Read off the quantity of N. 1.549 cubic inches of N at 60° Fah. and 30″ bar. = 1 grain urea.

Method inaccurate since ammonia, uric acid, &c., are likewise decomposed.

C. — *Heintz and Ragsky's Method.*

First determine ammonia by precipitation with PtCl₄.

Heat 2 to 5 c. c. with equal vol. H_2SO_4 in a covered capsule to 180°–200°. Cool, dilute with water, filter, and determine NH_3 formed by PtCl₄. Calculate both amounts for 100 c. c., and take the difference ; this multiplied by 0.13423 gives per cent. of urea.

Results very accurate.

D. — *Apjohn's Method.*

See "American Chemist," V. 431.

Provide the following apparatus :

(1) A glass tube 30 cm. long, subdivided into 30 equal parts, whose aggregate volume is 55 c. c. The end of the tube is drawn out like a Mohr's burette.

(2) A wide-mouthed gas bottle of 60 c. c. capacity.

(3) A test tube of 10 c. c. capacity, and long enough to be slightly inclined when introduced into the gas bottle.

The principle of the process is based upon the following equation :

$$2(CON_2H_4) + 3(CaBr_2O_2) = 3CaBr_2 + 2CO_2 + N_4$$

To make the hypobromite solution take 100 grms. NaHO, 250 c. c. H_2O, and add 25 c. c. bromine ; agitate and set aside for use.

Process : Into a glass cylinder containing water the tube (1) is depressed till the zero mark and surface of water coincide. 15 c. c. hypobromite solution (100 grms. NaHO, 250 c. c. H_2O, 25 c. c. Br) are placed in (2) and the test-tube containing the urine is introduced carefully to avoid spilling its contents. The flask is closed by a perforated stopper which is connected by tubing with the measuring tube. The urine is now mixed with the hypobromite, and the disengaged nitrogen is driven into the measuring tube. The tube is now levelled to relieve hydrostatic pressure, and the volume of nitrogen read off. Since 55 c. c. equal 0.15 grm. of urea, a single division corresponds to $\frac{0.15}{30}$ =0.005 grm. urea.

(0.15 grm. urea gives 55 c. c. nitrogen at 60° Fah. and 30° bar.)

7. Determination of Actual Ammonia. Take 20 c. c. filtered urine and treat by Schlösing's method.

The NH$_3$ is expelled by milk of lime, and absorbed by standard acid, in the cold under a bell jar. For details see *Fres.* § 99, 3 b. (Human urine contains 0.078 to 0.143 per cent.)

8. Determination of Albumen. Measure urine passed in 24 hours. Drop 50 c. c., one c. c. at a time, into 1 ounce boiling distilled water in a porcelain dish. If the urine was alkaline add a drop of acetic acid, avoid excess. Allow the coagulated albumen to settle, filter through a weighed filter, and wash well. Dry at 100° C, and weigh.

9. Determination of Sugar. Dilute urine 5 or 10 times, and apply Fehling's solution as in grape sugar. See *Analysis No.* 35, *Raw Sugar.*

10. Determination of Phosphoric Acid. To 50 c. c. filtered urine add 5 c. c. sodic acetate and titrate with uranic acetate. For details see Sutton's "Volumetric Analysis."

filter on a very small weighed filter. Wash-water should not exceed 30 c. c. If more is necessary add 0.045 mgm. uric acid for each c. c. additional. (Albumen must first be removed by coagulation.) Dry at 100°c. and weigh.

12. TESTS FOR BILE.

(1) Place a little urine on a white plate, add HNO_3. A peculiar play of colors — green, yellow, violet, &c. — occurs if coloring matter of bile is present.

(2) Agitate concentrated urine with boiling ether. If bile is present the ether solution will be greenish-yellow.

(3) Add baric acetate to urine, treat the ppt with alcohol, decompose it with HCl, and evaporate the liquid to dryness. Water will dissolve out in the residue coloring matter of the bile.

(4) *Pettenkofer's Test.*—Mix fluid with one-half vol. H_2SO_4, avoiding rise of temperature ; add a little powdered cane sugar ; mix and add more H_2SO_4. Liberation of cholalic acid produces a purplish-red coloration ; this gives a peculiar absorption spectrum. See Thudichum's "Manual."

Scheme for analysis of Urinary Sediments. (ATTFIELD.)

Warm the sediment with the supernatant urine, and filter.

INSOLUBLE.		SOLUBLE.
Phosphates, oxalate of calcium and uric acid. Warm with acetic acid, and filter.		Urates of Ca, Na, and NH_4,— chiefly of Na. They are re-deposited as the liquid cools, and if sufficient in quantity may be examined for uric acid and bases by usual tests.
INSOLUBLE. Oxalate of calcium and uric acid. Warm with HCl, and filter.	**SOLUBLE.** Phosphates. Add NH_4HO, and examine ppt. for P_2O_5, CaO and MgO.	
INSOLUBLE. Uric acid. Apply murexid test.	**SOLUBLE.** Oxalate of calcium. May be pptd. by NH_4HO.	

Note. — Urates are often of a pink or red color, owing to the pigment purpurine. This is soluble in alcohol.

Scheme for Determination of Urinary Sediments by Chemical Tests. (ATTFIELD.)

The sediment is white; warm with the supernatant urine and filter.		The sediment is colored		
		and crystalline *uric acid.*	and amorphous easily soluble on heating *urates.*	and amorphous, slowly soluble on heating. *Urates* colored by purpurine.
Solution contains *urates.*	Residue Treat with ammonia.			
	Solution contains *cystine.*	Residue Treat with acetic acid.		
	Residue *oxalate* and *oxalurate of calcium.*	Solution. Add NH₄ HO white ppt. of *earthy phosphates.*		

Analysis No. 34. — MILK.

A. — Determination of Water.

Wash quartz sand thoroughly with HCl and water, and ignite. Put about one-quarter inch of this sand in a platinum pan, weigh, and pour on 3 to 5 grms milk. Dry at 100° C. to constant weight.

B. — Determination of Butter.

Break up the cake from residue **A** and wash the butter out with ether into a weighed beaker, evaporate the ether and weigh the butter.

C. — Determination of Sugar.

Collect the residue from **B** on a dried and weighed filter, dry it at 100° C., boil it four or five times with fresh portions (150 c.c. each) of 80 per cent. alcohol, and dry the insoluble residue at 100° C. and weigh on a tared filter. The loss of weight gives the sugar approximately. Or determine sugar as under grape sugar, Analysis No. 35.

A convenient apparatus for the extraction of sugar is described by Prof. S. W. Johnson, in Am. J. of Sci. (3) xiii. page 196 (1877).

D. — Determination of total Non-volatile Matter.

Evaporate 10 to 20 grms. milk to dryness, with the addition of a little acetic acid, and ignite the residue in a muffle furnace, at the lowest possible temperature.

E. — Determination of Protein Compounds.

Subtract the sum of the butter, sugar, and ash from the total dry substance, and the remainder is chiefly casein.

For other methods, see "A Method for the Analysis of Milk," by E. H. von Baumhauer, *Am. Chem.*, Vol. VII., 191.

Analysis No. 35. — RAW SUGAR.
$$C_{12}H_{22}O_{11}$$

A. — Determination of Moisture.

Heat a weighed amount of sugar at 110° until it no longer loses in weight. Loss = moisture.

B. — Determination of Ash.

Weigh off ten grms. in a platinum dish. Either burn the sugar direct, or add a few drops of conc. H_2SO_4 and heat very cautiously in a gas muffle. Weigh the ash.

The two methods do not give results at all concordant; the latter is the French method, and the results are called "the salts," after subtracting one-ninth, but this is seldom correct, though the ash burns very white.

C. — Determination of Grape Sugar.
$$C_6H_{12}O_6, H_2O$$

(1) *Qualitative reactions.* Glucose is colored dark-

brown when heated with a strong solution of sodic hydrate. It dissolves in cold conc. H_2SO_4 without being
blackened. [Cane sugar blackens.]

If a conc. solution of glucose is mixed with cobaltic
nitrate, and a small quantity of fused NaHO, the solution
remains clear on being boiled ; if very concentrated it deposits a light-brown ppt.

[Cane sugar solutions similarly treated give a violet ppt.,
which turns green on standing].

BaH_2O_2 added to an alcoholic solution of glucose forms
a white ppt.

If a little caustic soda is added to a solution of glucose,
and then drop by drop a dilute solution of $CuSO_4$, a deep-
blue liquid forms ; after some time in the cold, but immediately if heated, a yellowish or red ppt. of hydrated cuprous
oxide is deposited. $\frac{1}{100.000}$ of glucose may be easily detected ; $\frac{1}{1.000.000}$ still gives a red tint to the solution.

Cupric acetate is similarly reduced. Potassio-tartrate of
copper acts likewise.

(2) *Quantitative estimation.* 1 eq. glucose will reduce
10 eq. of cupric oxide to cuprous oxide.

Preparation of Fehling's Solution.

Dissolve exactly 34.639 grms. pure dry $CuSO_4$ in about
200 c. c. water. In another vessel dissolve 173 grms. C. P.
Rochelle salts ($C_4H_4K NaO_6 + 4H_2O$) in 480 c. c. pure sodium hydrate solution having a sp. gr. 1.14.

Mix the solutions and dilute to exactly 1000 c. c. 10 c. c.
of this solution contains 0.34639 grms. $CuSO_4$ and corresponds to 0.050 grms. anhydrous glucose. Keep in the
dark. On boiling with four vols. of water, it should give
no precipitate.

The solution of glucose should not contain more than $\frac{1}{2}$ per cent. glucose ; if stronger, dilute.

Performance of Analysis :

Run exactly 10 c. c. of the copper solution into a small flask, add 40 c. c. water, (or a dilute solution of NaHO,) heat to boiling and run into the solution the liquid containing the glucose, slowly and gradually, from an accurate burette. Continue until the last shade of bluish green disappears, and a small portion of liquid filtered, gives no reaction with H_2S, nor with $HC_2H_3O_2$ and $K_4Fe_2Cy_6$.

Calculation. Since we took 10 c. c., Fehling's solution, corresponding to 0.050 grms. anhydrous glucose, we read off the number of c. c. of glucose solution taken ; this shows us how much of the substance contains 50 grms. grape sugar.

Example. — Used 9. 5 c. c. solution containing glucose :

$$9.5 : .05 = 100 : x$$

If solution was diluted, then $x \times d =$ per cent. glucose.

This method may be applied to cane sugar, by first converting it into grape sugar by boiling one to two hours with dilute H_2SO_4 (1 part acid 5 parts water). This is not very accurate, owing to formation of caramel. Milk sugar reduces Fehling's solution direct, but in another proportion, 100 glucose = 134 milk sugar.

D. — Determination of Crystalizable Cane Sugar.

Weigh out x grms.* of sugar or syrup, add water so that the whole will form about 80 c. c. Dissolve and add for

---

* The value of x depends upon the instrument employed. Instructions usually accompany a saccharimeter.

syrup 5 to 10 c. c. basic acetate of lead ; for raw sugar less ; for pure sugar, none. Dilute to 100 c. c. ; pour into a beaker, and add pulverized bone-black, and filter ; do not wash. Fill the tube of a Soleil or Dubosq Saccharimeter with this solution, perfectly full, insert the tube, and observe the transition tint. For details, see Atkinson's translation of Ganot's Physics, § 613. Cf. Fownes' Chemistry, p. 84, and Watts' Dict. iii. 673-5.

Analysis of a sample of RAW SUGAR.

Water, ...	2.07
Ash, ...	1.58
Grape Sugar, ...	1.82
Cane Sugar, ...	86.00

Analysis No. 37. — PETROLEUM.

For information as to the composition and refining of petroleum, the products which it yields by distillation, and the methods of testing kerosene, see Dr. C. F. Chandler's "Report on Petroleum Oil" in the "American Chemist," Vol. II. pp. 409, 446, and Vol. III. pp. 20 and 41.

A.—Distillation of Petroleum.

The method of examining crude petroleum for determination of its commercial value, is not that of fractional distillation in its true, scientific sense, but consists in a process of distillation which separates the liquid into a certain number of aliquot parts, having determinable densities, and flashing points ; and the value of the sample depends upon the proportion of the light and heavy products.

The process of distillation is conducted as follows. Select a tubulated retort of strong glass, free from flaws, and of about 500 c. c. capacity; connect this with a Liebig's condenser, and arrange for distilling in the usual manner. Through the tubulus of the retort insert a thermometer. Provide ten glass cylinders of 50 to 75 c. c. in capacity, and mark each with a file, so as to show the volume occupied by 25 c. c. of liquid. These cylinders are to serve as recipients of the distillate.

Pour 250 c. c. crude petroleum into the retort, and apply heat very gently at first, increasing gradually, and finally heating until the residue in the retort is coked. Collect 25 c. c. of the distillate in the first cylinder, and note the temperature indicated by the thermometer in the retort; collect the second 25 c. c. in another recipient, note also temperature, and continue in this manner, changing the recipient for every 25 c. c. until the whole liquid has distilled over.

B.—Examination of the Distillates.

Determine the sp. gr. of each distillate by floating in it a small Baumé Hydrometer, note the color of each sample, and determine its flashing point by means of Tagliabue's "Open Tester," a figure and description of which are found on page 41, Vol. III. of the "American Chemist."

To test the flashing point, proceed as follows : pour a small quantity of the sample to be examined into the open cup, which is surrounded by a vessel of water. Light the lamp beneath and apply heat very gradually ; the temperature should not rise faster than two degrees a minute. The thermometer bulb should dip beneath the surface of the oil. From time to time test the inflammable vapors which arise from the surface of the oil, using a small flame, flitting it quickly across the surface, and noting simultaneously the height of the thermometer at the moment of ignition. Record results with each distillate.

Example. — The following report of an actual distillation shows how the results may be reported.   This distillation was accompanied with the phenomena technically
called "cracking," by which the heavier hydrocarbons split
up into lighter ones.

No. of fraction.	Color.	Temperature Fahr.	Sp. Gr. Beaumé.	Flashing Point. Fahr.
1.	Colorless,	142°–224°	64	20°
2.	"	224 –298	60	48
3.	Light yellow,	298 –404	55	102
4.	"	404 –458	51	147
5.	"	458 –532	45	208
6.	Yellow,	532 – ?	42	254
7	Dark yellow,		40	204
8.	Deeper "		42	114
9.	Green,		44	82
10.	Black;			

The tenth product was coke left in the retort.

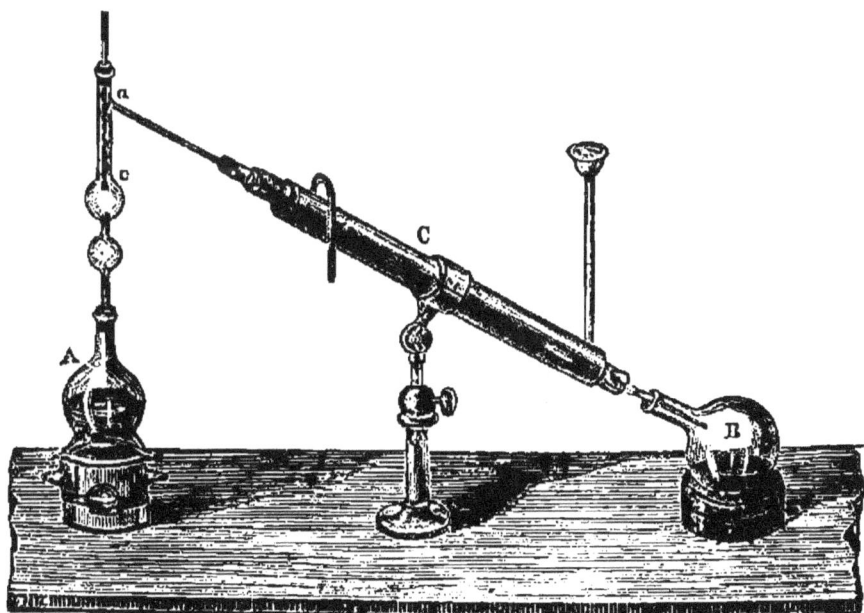

Fig. 6 shows the disposition of apparatus at the commencement of the distillation; so soon
as the lighter products have passed over, the bulb tube *a c* must be removed and connection
made with the condenser by a short bent tube.

TABLE I.

THE ELEMENTS, THEIR SYMBOLS, AND ATOMIC WEIGHTS.

Name.	Symbol.	Atomic Weight.	Name.	Symbol.	Atomic Weight.
Aluminium . . .	Al	27.0	Molybdenum . .	Mo	96.
Antimony . . .	Sb	120.	Nickel	Ni	58.8
Arsenic	As	75.	Nitrogen	N	14.
Barium	Ba	137	Osmium	Os	195.2
Bismuth	Bi	207.5	Oxygen	O	16.
Boron	B	11.	Palladium . . .	Pd	106.6
Bromine	Br	80.	Phosphorus . . .	P	31.
Cadmium	Cd	112.	Platinum	Pt	194.5
Caesium	Cs	133.	Potassium . . .	K	39.1
Calcium	Ca	40.	Rhodium . . .	Rh	104.4
Carbon	C	12.	Rubidium . . .	Rb	85.4
Cerium	Ce	141.	Ruthenium . . .	Ru	103.5
Chlorine	Cl	35.5	Samarium . . .	Sa	(?)
Chromium . . .	Cr	52.5	Scandium . . .	Sc	44.
Cobalt	Co	59.	Selenium	Se	79.4
Columbium . . .	Nb	94.	Silicon	Si	28.
Copper	Cu	63.4	Silver	Ag	108.
Didymium . . .	D	146.	Sodium	Na	23.
Erbium	E	166.	Strontium . . .	Sr	87.6
Fluorine	F	19.	Sulphur	S	32.
Gallium	Ga	70.	Tantalum . . .	Ta	182.
Glucinum	Be	9.	Tellurium . . .	Te	128.
Gold	Au	196.5	Terbium	Tb	(?)
Hydrogen . . .	H	1.	Thallium	Tl	204.
Indium	In	113.4	Thorium	Th	232.
Iodine	I	127.	Tin	Sn	118.
Iridium	Ir	192.5	Titanium . . .	Ti	50.
Iron	Fe	56.	Tungsten . . .	W	184.
Lanthanum . . .	La	138.5	Uranium	U	240.
Lead	Pb	206.5	Ytterbium . . .	Yb	173.
Lithium	Li	7.	Yttrium	Y	90.
Magnesium . . .	Mg	24.	Zinc	Zn	65.2
Manganese . . .	Mn	55.	Zirconium . . .	Zr	89.6
Mercury	Hg	200.			

TABLE II.

PRECIPITATING VALUE OF COMMON REAGENTS.

Solutions of reagents being prepared of the strength recommended by Fresenius (see Fres. Qual. Anal., § 17 to § 85, b, Johnson's edition of 1875), the amount of a reagent required for precipitation may be calculated from the following table :

One cubic centimetre of	Will precipitate
Dilute sulphuric acid	0.231 grm. Ba.
Barium chloride	0.032 " SO_3.
Hydrodisodic phosphate	0.011 " MgO.
Magnesia mixture	0.024 " P_2O_5
Ammonium molybdate	0.001 " P_2O_5.
Ammonium oxalate	0.016 " CaO.
Argentic nitrate	0.010 " Cl.

TABLE III.

DIAMETER OF FILTERS AND WEIGHTS OF FILTER ASHES; SWEDISH PAPER.

Filter No.	Diameter.	Weight of Ash.	
		Acid.	Alkaline.
1 . . .	70 mm.	0.0004 grm.	0.0014
2 . . .	104 "	0.0007 "	0.0027
3 . . .	122 "	0.0011 "	0.0043
4 . . .	147 "	0.0016 "	0.0062

TRINITY COLLEGE.

HARTFORD,..188 .

Report of

Analysis of

Determination of

Grammes taken :

Method of Analysis.

Precipitates.	Actual Weights.	Constituents.	Calculated Weights.	Percentages.	Theoretical Percentages.

Special Remarks.

[This is a reduced fac-simile of the reporting blank, measuring 8 by 10 inches, described on page 17.]

OFFICIAL METHODS OF ANALYSIS

OF THE ASSOCIATION OF OFFICIAL AGRICULTURAL CHEMISTS FOR 1887-'88.

METHODS FOR DETERMINING PHOSPHORIC ACID AND MOISTURE.

(1) *Preparation of Sample.*—The sample should be well intermixed and properly prepared, so that separate portions shall accurately represent the substance under examination, without loss or gain of moisture.

(2) *Determination of Moisture.*—(*a*) In potash salts, nitrate of soda, and sulphate of ammonia heat 1 to 5 grams at 130° C. till the weight is constant, and reckon water from the loss. (*b*) In all other fertilizers heat 2 grams, or if the sample is too coarse to secure uniform lots of 2 grams each, 5 grams for five hours at 100° in a steam bath.

(3) *Water-soluble Phosphoric Acid.*—Bring 2 grams on a filter, add a little water, let it run out before adding more water, and repeat this treatment cautiously until no phosphate is likely to precipitate in the filter. If the washings show turbidity after passing the filter clear up with acid. When the substance is nearly washed in this manner it is transferred to a mortar and rubbed with a rubber-tipped pestle to a homogeneous paste (but not further pulverized), then returned to the filter and washed with water until the filtrate measures not less than 250 cc. Mix the washings. Take an aliquot (usually corresponding to $\frac{1}{2}$ or $\frac{1}{4}$ gram of the substance) and determine phosphoric acid, as under total phosphoric acid.

(4) *Citrate-insoluble Phosphoric Acid.*—Wash the residue of the treatment with water into a 200 cc. flask with 100 cc. of strictly neutral ammonium citrate solution of 1.09 density, prepared as hereafter directed. Cork the flask securely and place it in a water bath, the water of which stands at 65° C. (The water bath should be of such a size that the introduction of the cold flask or flasks shall not cause a reduction of the temperature of the bath of more than 2° C.) Raising the temperature as rapidly as practicable to 65° C., which is

subsequently maintained, digest with frequent shakings for thirty
minutes from the instant of insertion, filter the warm solution quickly
(best with filter-pump), and wash with water of ordinary temperature.
Transfer the filter and its contents to a capsule, ignite until the
organic matter is destroyed, treat with 10 to 15 cc. of concentrated
hydrochloric or nitric acid, digest over a low flame until the phosphate is dissolved, dilute to 200 cc., mix, pass through a dry filter,
take an aliquot and determine phosphoric acid as under total.

In case a determination of citrate-insoluble phosphoric acid is required in non-acidulated goods, it is to be made by treating 2 grams
of the phosphatic material, without previous washing with water,
precisely in the way above described, except that in case the substance contains much animal matter (bone, fish, etc.) the residue
insoluble in ammonium citrate is to be treated by one of the pro
cesses described below.

(5) *Total Phosphoric Acid.*—Weigh 2 grams and treat by one of
the following methods : (1) Evaporation with 5 cc. magnesium
nitrate, ignition, and solution in acid. (2) Solution in 30 cc. concentrated nitric acid with a small quantity of hydrochloric acid. (3) Add
30 cc. concentrated hydrochloric acid, heat, and add cautiously and
in small quantities at a time about 0.5 gram of finely pulverized
potassium chlorate.

Boil gently until all phosphates are dissolved and all organic matter
destroyed ; dilute to 200 cc.; mix and pass through a dry filter ; take
50 cc: of filtrate ; neutralize with ammonia (in case hydrochloric acid
has been used as a solvent add about 15 grams dry ammonium
nitrate or its equivalent). To the hot solutions for every decigram
of P_2O_5 that is present, add 50 cc. of molybdic solution. Digest at
about 65° C. for one hour, filter, and wash with ammonium nitrate
solution. (Test the filtrate by renewed digestion and addition of
more molybdic solution.) Dissolve the precipitate on the filter with
ammonia and hot water and wash into a beaker to a bulk of not
more than 100 cc. Nearly neutralize with hydrochloric acid, cool,
and add magnesia mixture from a burette ; add slowly (one drop per
second), stirring vigorously. After fifteen minutes add 30 cc. of
ammonia solution of density 0.96. Let stand several hours (two
hours is usually enough). Filter, wash with dilute ammonia, ignite
intensely for ten minutes, and weigh.

(6) Citrate-soluble phosphoric acid. The sum of the water-soluble
and citric insoluble subtracted from the total gives the citrate-soluble.

6*

(1) *To Prepare Ammonium Citrate Solution.*—Mix 370 grams of commercial citric acid with 1,500 cc. of water; nearly neutralize with crushed commercial carbonate of ammonia; heat to expel the carbonic acid; cool; add ammonia until exactly neutral (testing by saturated alcoholic solution of coralline) and bring to volume of two liters. Test the specific gravity, which should be 1.09 at 20° C., before using.

(2) *To Prepare Molybdic Solution.*—Dissolve 100 grams of molybdic acid in 400 grams or 417 cc. of ammonia of specific gravity 0.96, and pour the solution thus obtained into 1,500 grams or 1,250 cc. of nitric acid of specific gravity 1.20. Keep the mixture in a warm place for several days, or until a portion heated to 40° C. deposits no yellow precipitate of ammonium phospho-molybdate. Decant the solution from any sediment, and preserve in glass-stoppered vessels.

(3) *To Prepare Ammonium Nitrate Solution.*—Dissolve 200 grams of commercial ammonium nitrate in water and bring to a volume of two liters.

(4) *To Prepare Magnesia Mixture.*—Dissolve 22 grams of recently ignited calcined magnesia in dilute hydrochloric acid, avoiding excess of the latter. Add a little calcined magnesia in excess, and boil a few minutes to precipitate iron, alumina, and phosphoric acid, filter, add 280 grams of ammonium chloride, 700 cc. of ammonia of specific gravity 0.96, and water enough to make the volume of two liters. Instead of the solution of 22 grams of calcined magnesia 110 grams of crystallized magnesium chloride ($MgCl_2$, $6H_2O$) may be used.

(5) *Dilute Ammonia for Washing.*—One volume ammonia of specific gravity 0.96 mixed with three volumes of water, or usually one volume of concentrated ammonia with six volumes of water.

(6) *Nitrate of Magnesia.*—Dissolve 320 grams of calcined magnesia in nitric acid, avoiding an excess of the latter; then add a little calcined magnesia in excess, boil; filter from excess of magnesia, ferric oxide, etc., and bring to volume of two liters.

METHODS OF DETERMINING POTASH.

METHOD OF LINDO AS MODIFIED BY GLADDING.

(1) *Superphosphates.*—Boil 10 grams of the fertilizer with 300 cc. of water for ten minutes. Cool the solution; add ammonia in slight

excess, thus precipitating all phosphate of lime, oxide of iron, and alumina, etc., make up to 500 cc., mix thoroughly and filter through a dry filter ; take 50 cc. corresponding to 1 gram, evaporate nearly to dryness, add 1 cc. of dilute H_2SO_4 (1 to 1), and evaporate to dryness and ignite to whiteness. As all the potash is in form of sulphate, no loss need be apprehended by volatilization of potash, and a full red heat must be used until the residue is perfectly white. This residue is dissolved in hot water plus a few drops of HCl ; 5 cc. of a solution of pure NaCl (containing 20 grams NaCl to the liter) and an excess of platinum solution (4 cc.) are now added, and the whole evaporated as usual. The precipitate is washed thoroughly with alcohol by decantation and on filter, as usual. The washing should be continued even after the filtrate is colorless. Ten cc. of the NH_4Cl solution prepared as above are now run through the filter. These 10 cc. will contain the bulk of the impurities, and are thrown away. A fresh portion of 10 cc. NH_4Cl is now run through the filter several times (five or six). The filter is then washed thoroughly with pure alcohol, dried, and weighed as usual. The platinum solution used contains 1 gram metallic platinum in every 10 cc.

(2) *Muriates of Potash.*—In the analysis of these salts an aliquot portion, containing .500 gram is evaporated with 10 cc. platinum solution plus a few drops of HCl, and washed as before.

(3) *Sulphate of Potash, Kainite, &c.*—In the analysis of these salts an aliquot portion containing .500 gram is taken, .250 gram of NaCl added, plus a few drops of HCl, and the whole evaporated with 15 cc. platinum solution. In this case special care must be taken, in the washing with alcohol, to remove all the double chloride of platinum and sodium. The washing should be continued for some time after the filtrate is colorless. Twenty-five cubic centimeters of the NH_4Cl solution are employed, instead of 10 cc., and the 25 cc. poured through at least six times to remove all sulphates and chlorides. Wash finally with alcohol, dry and weigh as usual.

To prepare the washing solution of NH_4Cl, place in a bottle 500 cc. H_2O, 100 grams of NH_4Cl ; shake till disolved. Now pulverize 5 or 10 grams of K_2PtCl_6, put in the bottle, and shake at intervals for six or eight hours ; let settle over night ; then filter off liquid into a second bottle. The first bottle is then ready for a preparation of a fresh supply when needed.

ALTERNATE METHOD.

Pulverize the fertilizer (200 or 300 grams) in a mortar; take 10 grams, boil for ten minutes with 200 cc. water, and after cooling, and without filtering, make up to 1,000 cc., and filter through a dry paper. In this method, in case the potash is contained in organic compounds, like tobacco stems, cottonseed hulls, &c., the substance is to be saturated with strong sulphuric acid and ignited in a muffle to destroy organic matter. If the sample have 10 to 15 per cent K_2O (kainite), take 50 cc. of the filtrate; if from 2 to 3 per cent K_2O (ordinary potash fertilizers), take 100 cc. of the filtrate. In each case make the volume up to 150 cc., heat to 100°, and add, drop by drop, with constant stirring, slight excess of barium chloride; without filtering, in the same manner, add barium hydrate in slight excess. Heat, filter, and wash until precipitate is free of chlorides. Add to filtrate 1 cc. strong ammonium hydrate, and then a saturated solution of ammonium carbonate until excess of barium is precipitated. Heat. Add now, in fine powder, 0.5 gram pure oxalic acid or 0.75 gram ammonium oxalate. Filter, wash free of chlorides, evaporate filtrate to dryness in a platinum dish, and, holding dish with crucible tongs, ignite carefully over the free flame below red heat until all volatile matter is driven off.

The residue is now digested with hot water, filtered through a small filter, and washed with successive small portions of water until the filtrate amounts to 30 cc. or more. To this filtrate, after adding two drops of strong hydrochloric acid, is added, in a porcelain dish, 5 to 10 cc. of a solution of 10 grams of platinic chloride in 100 cc. of water. The mixture is now evaporated on the water-bath to a thick syrup, or further treated with strong alcohol washed by decantation, collected in a Gooch crucible or other form of filter, washed with strong alcohol, afterwards with 5 cc. ether, dried for thirty minutes at 100° C., and weighed.

It is desirable, if there is an appearance of white foreign matter in the double salt, that it should be washed, according to the previous method, with 10 cc. of the half-concentrated solution of NH_4Cl, which has been saturated by shaking with K_2PtCl_6, as recommended by Gladding.

The use of the factor 0.3056 for converting K_2PtCl_6 to KCl and 0.19308 for converting to K_2O is continued.

METHOD FOR THE DETERMINATION OF NITROGEN.

THE ABSOLUTE OR CUPRIC OXIDE METHOD.

The apparatus and reagents needed are as follows:

APPARATUS.

Combustion Tube of best hard Bohemian glass, about 26 inches long and one-half inch internal diameter.

Azotometer of at least 100 cubic centimeters capacity, accurately calibrated.

Sprengel Mercury Air Pump.

Small Paper Scoop, easily made from stiff writing-paper.

REAGENTS.

Cupric Oxide (coarse).—Wire form ; to be ignited and cooled before using.

Fine Cupric Oxide.—Prepared by pounding ordinary cupric oxide in mortar.

Metallic Copper.—Granulated copper or fine copper gauze reduced and cooled in steam of hydrogen.

Sodium Bicarbonate.—Free from organic matter.

Caustic Potash Solution.—Dissolve commercial stick potash in less than its weight of water so that crystals are deposited on cooling. When absorption of carbonic acid ceases to be prompt solution must be discarded.

LOADING TUBE.

Of ordinary commercial fertilizers take 1 to 2 grams for analysis. In the case of highly nitrogenous substances the amount to be taken must be regulated by the amount of nitrogen estimated to be present.   Fill tube as follows : (1) About 2 inches of coarse cupric oxide. (2) Place on the small paper scoop enough of the fine cupric oxide to fill, after having been mixed with the substance to be analyzed, about 4 inches of the tube ; pour on this the substance, rinsing watch glass with a little of the fine oxide and mix thoroughly with spatula ; pour into tube, rinsing the scoop with a little fine oxide. (3) About 12 inches of coarse cupric oxide. (4) About 3 inches of metallic copper.

(5) About 2½ inches of coarse cupric oxide (anterior layer). (6) Small plug of asbestos. (7) Eight-tenths to 1 gram of sodium bicarbonate. (8) Large, loose plug of asbestos ; place tube in furnace, leaving about one inch of it projecting; connect with pump by rubber stopper smeared with glycerine, taking care to make connection perfectly tight.

OPERATION.

Exhaust air from tube by means of pump. When a vacuum has been obtained, allow flow of mercury to continue, light gas under that part of tube containing metallic copper, anterior layer of cupric oxide (see 5th above) and bicarbonate of soda. As soon as vacuum is destroyed and apparatus filled with carbonic acid gas, shut off the flow of mercury, and at once introduce the delivery tube of the pump into the receiving arm of the azotometer and just below the surface of the mercury seal of the azotometer, so that the escaping bubbles will pass into the air and not into the azotometer, thus avoiding the useless saturation of the caustic potash solution.

When the flow of carbonic acid has very nearly or completely ceased, pass the delivery tube down into the receiving arm, so that the bubbles will escape into the azotometer. Light the jets under the 12-inch layer of oxide, heat gently for a few moments to drive out any moisture that may be present, and bring to red heat. Heat gradually mixture of substance and oxide, lighting one jet at a time. Avoid too rapid evolution of bubbles, which should be allowed to escape at rate of about one per second or a little faster.

When the jets under mixture have all been turned on, light jets under layer of oxide at end of tube. When evolution of gas has ceased turn out all the lights except those under the metallic copper and anterior layer of oxide, and allow to cool for a few moments. Exhaust with pump and remove azotometer before flow of mercury is stopped. Break connection of tube with pump, stop flow of mercury, and extinguish lights. Allow azotometer to stand at least an hour or cool with stream of water until permanent volume and temperature are reached.

Adjust accurately the level of the KOH solution in bulb to that in azotometer, note volume of gas, temperature, and height of barometer ; make calculations as usual. The labor of calculation may be much diminished by the use of the tables prepared by Messrs.

Battle and Dancy, of the North Carolina Experiment Station (Raleigh, N. C.).

The above details are, with some modifications, those given in the report of the Connecticut Station for 1879 (p. 124), which may be consulted for details of apparatus, should such details be desired.

DETERMINATION BY THE METHOD OF KJELDAHL.

REAGENTS AND APPARATUS.

(1) Hydrochloric acid whose absolute strength has been determined, (*a*) by precipitating with silver nitrate and weighing the silver chloride, (*b*) by sodium carbonate, as described in Fresenius's Quantitative Analysis, second American edition, page 680, and (*c*) by determining the amount neutralized by the distillate from a weighed quantity of pure ammonium chloride boiled with an excess of sodium hydrate.

(2) Standard ammonia whose strength, relative to the acid, has been accurately determined.

(3) " C. P." sulphuric acid, specific gravity 1.83, free from nitrates and also from ammonium sulphate, which is sometimes added in the process of manufacture to destroy oxides of nitrogen.

(4) Mercuric oxide, HgO, prepared in the wet way. That prepared from mercury nitrate cannot safely be used.

(5) Potassium permanganate tolerably finely pulverized.

(6) Granulated zinc.

(7) A solution of 40 grams of commercial potassium sulphide in one liter of water.

(8) A saturated solution of sodium hydrate free from nitrates, which are sometimes added in the process of manufacture to destroy organic matter and improve the color of the product. That of the Greenbank Alkali Company is of good quality.

(9) Solution of cochineal prepared according to Fresenius's Quantitative Analysis, second American edition, page 679.

(10) Burettes should, be calibrated in all cases by the user.

(11) Digestion flasks of hard, moderately thick, well-annealed glass. These flasks are about 9 inches long, with a round, pear-shaped bottom, having a maximum diameter of 2½ inches, and tapering out gradually in a long neck, which is three-fourths of an inch in diam-

eter at the narrowest part, and flared a little at the edge.   The
total capacity is 225 to 250 cc.

(12) Distillation flasks of ordinary shape, 550 cc. capacity, and
fitted with a rubber stopper and a bulb tube above to prevent the
possibility of sodium hydrate being carried over mechanically during
distillation.   This is adjusted to the tube of the condenser by a rubber tube.

(13) A condenser.   Several forms have been described, no one of
which is equally convenient for all laboratories.   The essential thing
is that the tube which carries the steam to be condensed shall be of
block tin.   All kinds of glass are decomposed by steam and ammonia
vapor, and will give up alkali enough to impair accuracy.   (See Kreus-
sler and Henzold, *Ber. d. chem. Ges.*, *XVII.*, 34.)   The condenser in
use in the laboratory of the Connecticut Experiment Station, devised by
Professor Johnson, consists of a copper tank supported by a wooden
frame, so that its bottom is 11 inches above the work-bench on which
it stands.   This tank is 16 inches high, 32 inches long, and 3 inches
wide from front to back, widening above to 6 inches.   It is provided
with a water-supply tube which goes to the bottom, and a larger
overflow pipe above.   The block tin condensing tubes, whose exter-
nal diameter is $\frac{3}{8}$ of an inch, seven in number, enter the tank through
holes in the front side of it near the top, above the level of the overflow, and pass down perpendicularly through the tank and out
through rubber stoppers, tightly fitted into holes in the bottom.   They
project about 1½ inches below the bottom of the tank, and are connected by short rubber tubes with glass bulb tubes of the usual shape,
which dip into glass precipitating beakers.   These beakers are
6½ inches high, 3 inches in diameter below, somewhat narrower
above, and of about 500 cc. capacity.   The titration can be made
directly in them.   The seven distillation flasks are supported on a
sheet-iron shelf attached to the wooden frame that supports the tank
in front of the latter.   Where each flask is to stand a circular hole
is cut, with three projecting lips, which support the wire gauze under
the flask, and three other lips which hold the flask in place and prevent its moving laterally out of place while distillation is going on.
Below this sheet-iron shelf is a metal tube carrying seven Bunsen
burners, each with a stop-cock like those of a gas-combustion furnace.   These burners are of larger diameter at the top, which prevents smoking when covered with fine gauze to prevent the flame
from striking back.

(14) The stand for holding the digestion flasks consists of a pan of sheet-iron 29 inches long by 8 inches wide, on the front of which is fastened a shelf of sheet-iron as long as the pan, 5 inches wide and 4 inches high. In this are cut six holes 1⅛ inches in diameter. At the back of the pan is a stout wire running lengthwise of the stand, 8 inches high, with a bend or depression opposite each hole in the shelf. The digestion flask rests with its lower part over a hole in the shelf and its neck in one of the depressions in the wire frame, which holds it securely in position. Heat is supplied by low Bunsen burners below the shelf. With a little care the naked flame can be applied directly to the flask without danger.

THE DETERMINATION.

One gram of the substance to be analyzed is brought into a digestion flask with approximately 0.7 gram of mercuric oxide and 20 cc. of sulphuric acid. The flask is placed on the frame above described in an inclined position and heated below the boiling point of the acid for from 5 to 15 minutes, or until frothing has ceased. The heat is then raised till the acid boils briskly. No further attention is required till the contents of the flask have become a clear liquid, which is colorless or at least has only a very pale straw color. The flask is then removed from the frame, held upright, and, while still hot, potassium permanganate is dropped in carefully and in small quantity at a time till after shaking the liquid remains of a green or purple color. After cooling, the contents of the flask are transferred to the distilling flask with water, and to this 25 cc. of potassium sulphide solution are added, 50 cc. of the soda solution, or sufficient to make the reaction strongly alkaline, and a few pieces of granulated zinc. The flask is at once connected with the condenser and the contents of the flask are distilled till all ammonia has passed over into the standard acid contained in the precipitating flask previously described and the concentrated solution can no longer be safely boiled. This operation usually requires from 20 to 40 minutes. The distillate is then titrated with standard ammonia.

The use of mercuric oxide in this operation greatly shortens the time necessary for digestion, which is rarely over an hour and a half in the case of substances most difficult to oxidize and is more commonly less than an hour. In most cases the use of potassium permanganate is quite unnecessary, but it is believed that in exceptional cases it is required for complete oxidation, and in view of the uncertainty it is always used. Potassium sulphide removes all mercury from solution and so prevents the formation of mercuro-ammonium compounds which are not completely decomposed by soda solution. The addition of zinc gives rise to an evolution of hydrogen and prevents violent bumping. Previous to use, the reagents should be tested by a blank experiment with sugar, which will partially reduce any nitrates that are present which might otherwise escape notice.

The following modification must be used for the determination of nitrogen in substances which contain nitrates when it is desired to use this method :

DETERMINATION OF NITROGEN, INCLUDING THE NITROGEN OF NITRATES, BY A MODIFIED METHOD OF KJELDAHL.[a]

Bring from 07. to 1.4 grams of the substance to be analyzed into a Kjeldahl digesting flask, add to this 30 cc. of sulphuric acid containing 2 grams of salicylic acid, and shake thoroughly. Then add *gradually* three grams of zinc-dust, shaking the contents of the flask at the same time. Finally, add two or three drops of platinic chloride solution and place the flask on the stand for holding the digestion flasks, where it is heated over a low flame until all danger from frothing has passed. The heat is then raised until the acid boils briskly, and the boiling continued until white fumes no longer pour out of the flask. This requires about five or ten minutes. Add now approximately 0.7 gram mercuric oxide and continue the boiling until the liquid in the flask is colorless, or nearly so. (In case the contents of the flask are likely to become solid before this point is reached add 10 cc. more of sulphuric acid.) Complete this oxidation with a little permanganate of potash in the usual way, and proceed with the distillation as described in the method above.

DETERMINATION BY THE RUFFLE METHOD.

PREPARATION OF REAGENTS.

(1) *A standard solution of sulphuric acid*, half normal, or 19.968 grams SO_3 per liter.

(2) *A standard solution of potassium hydrate*, half normal, or 27.991 grams KOH per liter.

---

[a]Described by Prof. M. A. Scovell.

(3) *An alcoholic solution of cochineal.*

(4) *Hyposulphite mixture.*—Prepared by mixing equal parts by weight of soda-lime and finely powdered crystallized sodium hyposulphite.

(5) *Sugar and sulphur mixture.*—This is prepared by mixing equal parts by. weight of finely powered granulated sugar and flowers of sulphur.

(6) *Ordinary granulated soda-lime.*

APPARATUS.

(1) *Combustion tubes* of hard Bohemian glass, 20 inches long and ¼ inch internal diameter, drawn to a point.

(2) *Three-bulb, 6-inch U tubes,* with glass stop-cock.

(3) *Aspirator.*

PREPARATION.

(1) Clean and fill the U tube with 10 cubic centimeters of standard acid,

(2) Fit cork and glass connecting tube. Fill the tube as follows : (1) A loosely fitting plug of asbestos, previously ignited, and then 1 to 1¼ inches of the hyposulphite mixture. (2) The weighed portion of the substance to be analyzed is intimately mixed with from 5 to 10 grams of the sugar and sulphur mixture. (3) Pour on a piece of glazed paper, or porcelain mortar, a sufficient quantity of the hyposulphite mixture to fill about 10 inches of the tube ; then add the substance to be analyzed, as previously prepared ; mix carefully and pour into the tube ; shake down the contents of the tube ; rinse off the paper or mortar with a small quantity of the hyposulphite mixture and pour into the tube ; then fill up with soda-lime to within 2 inches of the end of the tube. (4) Place another plug of ignited asbestos at the end of the tube, and close with a cork. (5) Hold the tube in a horizontal position, and tap on the table until there is a gas channel all along the top of the tube. Make connection with the U tube containing the acid, aspirate, and see that the apparatus is tight.

The Combustion.—Place the prepared combustion tube in the furnace, letting the open end project a little so as not to burn the cork. Commence by heating the soda-lime portion until it is brought to a full red heat. Then turn up slowly jet after jet towards the outer end of the tube, so that the bubbles come off two or three a

second. When the whole tube is red hot and the evolution of the gas
has ceased and the liquid in the U tube begins to recede toward the
furnace, attach the aspirator to the other limb of the U tube, break
off the end of the tube, and draw a current of air through for a few
minutes. Detach the U tube and wash the contents into a beaker
or porcelain basin, add a few drops of the cochineal solution, and
titrate.

DETERMINATION BY THE SODA-LIME METHOD.

Select a tube of hard glass 14 to 16 inches long, draw one end of it
to a fine point, and to the other end fit a cork, through which is passed
a tube bent at right angles, the other end of which passes through a
cork closing tightly one arm of a 6-inch three-bulbed U tube with
glass stop-cock. Into the combustion tube first slip a loosely fitting
plug of asbestos previously ignited, and then $1\frac{1}{2}$ to 2 inches of soda-
lime. Weigh out from 0.7 to 2.8 grams of the substance to be
analyzed, and mix it on a piece of glazed paper or porcelain mortar with some finely pulverized soda-lime, and introduce the mixture into the combustion tube. The paper or mortar is then rinsed
out with a small quantity of soda-lime and poured into the tube.
The tube is then filled up to within 1 to 2 inches of the open end
with granulated soda-lime, then with a plug of ignited asbestos.
A free passage is formed for the evolved gases by holding the
tube in a horizontal position and tapping gently on the table. Introduce the prepared combustion tube into the furnace, letting the
open end project a little so as not to burn the cork, supporting
the U tube by a clamp. The tube is then gradually heated, com-
mencing at the fore part, nearest the cork, and progressing slowly
towards the tail. The combustion should be conducted so as to
obtain a steady and uninterrupted flow of gas. When properly
carried out, the acid in the U tube is never colored. When the
acid begins to recede attach the aspirator to the other limb of
the U tube, and start it slowly, then break off the point of the
combustion tube and draw a current of air through the apparatus
for a few minutes, in order to sweep all the ammonia into the acid.

Detach the U tube and wash the contents into a beaker or
porcelain basin, add a few drops of an alcoholic solution of cochineal, and titrate.

The standard acid and alkali to be the same as that used in
the Ruffle method.

141

APPENDIX.

www.ingramcontent.com/pod-product-compliance
Lightning Source LLC
Chambersburg PA
CBHW021817190326
41518CB00007B/632